Geology of the Carrickfergus area

The area described in this memoir lies to either side of Belfast Lough, which separates the generally older rocks in County Down from the younger strata and volcanic rocks of County Antrim to the north. Although the area has been dealt with in part in several earlier memoirs this is the first comprehensive account of the geology of the area covered by the published one-inch geological map.

Following an introductory chapter which outlines, among other things, previous geological work in the area; the stratigraphy, palaeontology, petrography and structure of the Lower Palaeozoic rocks are described. There are detailed accounts of the Lower Carboniferous, Permian, Triassic, Jurassic and Cretaceous strata together with a description of the Tertiary volcanic rocks. The latter includes a detailed description of the geology of the Carneal Plug, discovered during the resurvey, and which has yielded such a spectacular mineralogy.

The complex structure of the Lower Palaeozoic rocks is described, as are the structures attributable to later orogenic movements.

The Pleistocene and Recent deposits, which influence building and engineering projects throughout the area, are discussed.

The possible extension of the Southern Uplands Fault into Northern Ireland is discussed in the chapter dealing with geophysical investigations and there are sections on the history of lead and salt mining. The prospects for finding buried coal and hydrocarbons in south-east Antrim are also considered.

In order to keep the price of this memoir to a minimum but at the same time to make available the details of the numerous sections and boreholes which are recorded and to publish the palaeontological records five Appendices have been prepared as microfiche plates.

Carrickfergus Castle: A Norman stronghold situated on a massive Tertiary dolerite dyke and built mostly of dolerite and basalt blocks with coigns of pale cream-coloured Permian limestone, probably from Cultra [NI 376]
Frontispiece

GEOLOGICAL SURVEY OF NORTHERN IRELAND
Department of Economic Development

A. E. GRIFFITH and
H. E. WILSON

Geology of the country around Carrickfergus and Bangor

Memoir for one-inch geological sheet 29

SECOND EDITION

CONTRIBUTORS

J. R. P. Bennett, A. Brandon,
E. J. Cobbing, T. P. Fletcher,
R. R. Harding, J. R. Hawkes,
H. C. Ivimey-Cook, P. I. Manning,
M. Mitchell, R. A. Old,
J. E. Robinson, A. W. A. Rushton,
P. A. Sabine, R. W. Sanderson,
C. J. Wood and A. Woodrow

INSTITUTE OF GEOLOGICAL SCIENCES
Natural Environment Research Council

BELFAST HER MAJESTY'S STATIONERY OFFICE 1982

iv

© *Crown Copyright 1982*

Bibliographical reference

GRIFFITH, A. E. and WILSON, H. E. 1982. Geology of the country around Carrickfergus and Bangor. *Mem. Geol. Surv. North. Irel.*, Sheet 29.

Authors

A. E. GRIFFITH, BSc
Geological Survey of Northern Ireland,
Belfast

H. E. WILSON, MSc, MIMM, MIGeol
Institute of Geological Sciences,
Keyworth, Nottingham NG12 5GG

Contributors

J. R. P. Bennett, BSc
Geological Survey of Northern Ireland,
Belfast

R. R. Harding, BSc, DPhil, FGA, J. R. Hawkes, BSc, PhD, H. C. Ivimey-Cook, BSc, PhD, A. W. A. Rushton, BA, PhD, P. A. Sabine, DSc, PhD, ARCS, FRSA, FRSE, FFMM, R. W. Sanderson, BSc and C. J. Wood, BSc
Institute of Geological Sciences, London

A. Brandon, BSc, PhD., E. J. Cobbing, BSc, PhD and P. I. Manning, TD, BSc, MIWES
Institute of Geological Sciences,
Keyworth, Nottingham NG12 5GG

T. P. Fletcher, MSc, PhD and M. Mitchell, MA
Institute of Geological Sciences,
Ring Road Halton, Leeds LS15 8TQ

Printed in Northern Ireland for Her Majesty's Stationery Office.
Dd.014274. C20. 3/83.
ISBN 0 11 884343 5

Other publications of the Geological Survey of Northern Ireland

Books

The composition and origin of the Antrim laterites
 and bauxites
The geology of the country around Dungannon
The geology of the country around Ballycastle
The geology of Belfast and the Lagan Valley
Geology of the Causeway Coast Vols. 1 and 2
The regional geology of Northern Ireland

Obtainable from Government Bookshops in Belfast, London, Edinburgh, Cardiff, Manchester, Birmingham and Bristol, or booksellers. Post orders to Government Bookshop, Chichester Street, Belfast, BT1 4JY

Geological maps

Six inches to one mile (1:10 560) maps of coalfield areas:
 Antrim: 5 SW, 5 SE, 9 NW: Tyrone: 46 SE, 47 NW,
 47 SW, 54 NE
Three inches to one mile (1:21 120) map:
 Engineering geology of the Belfast district

One inch to one mile (1:63 360) colour-printed maps:
 Giant's Causeway (7) Sheet (Solid)
 Giant's Causeway (7) Sheet (Drift)
 Ballycastle (8) Sheet (Solid)
 Ballycastle (8) Sheet (Drift)
 Limavady (6/12) Sheet (Solid)
 Limavady (6/12) Sheet (Drift)
 Cookstown (27) Sheet (Solid)   (*in press*)
 Cookstown (27) Sheet (Drift)   (*in press*)
 Carrickfergus (29) Sheet (Solid)
 Carrickfergus (29) Sheet (Drift)
 Pomeroy (34) Sheet (Solid)
 Pomeroy (34) Sheet (Drift)
 Dungannon (35) Sheet (Solid)
 Dungannon (35) Sheet (Drift)
 Enniskillen (45) Sheet (Solid)
 Enniskillen (45) Sheet (Drift)
 Mourne Mountains Special Sheet (Solid)
1:250 000 map:
 Northern Ireland

Geophysical maps

Quarter-inch to one mile (1:253 440) maps:
 Gravity anomaly map of Northern Ireland
 Aeromagnetic map of Northern Ireland

Obtainable from Ordnance Survey, Ladas Drive, Belfast BT7 9FJ, or Ordnance Survey agents

A full list of provisional maps and open-file reports may be obtained from the Geological Survey of Northern Ireland, 20 College Gardens, Belfast, BT2 6BS

CONTENTS

MICROFICHE CONTENTS

Appendix 1 Records of coastal sections of Permian and lower Carboniferous rocks at Cultra, Co. Down
Appendix 2 List of Lower Carboniferous fossils from Cultra shore sections
Appendix 3 Notes on certain ostracod species
Appendix 4 Borehole and shaft records
Appendix 5 Details of Ordovician, Silurian, Carboniferous, Triassic, Tertiary: Antrim Lava Group, Intrusive igneous rocks and Pleistocene exposures

PLATES

FIGURES

TABLES

PREFACE TO THE SECOND EDITION

The district covered by the Carrickfergus (29) Sheet of the one-inch to one mile geological map of Ireland was first surveyed in 1867–8 by G. V. Du Noyer, who mapped the whole area except for a small area in the south-east. This he 'corrected' after it had been covered by W. B. Leonard, who was presumably under training as he joined the Geological Survey in 1867. After Du Noyer's death in 1869 E. Hull and J. W. Warren 'partly revised' the County Down area and the first edition of the map was published in 1869. A Memoir covering the area to the south of Belfast Lough, and including a description of Sheets 37 and 38 was published in 1871. Another Memoir, covering the area to the north, and Sheets 21 and 28 was published in 1876.

In 1882, or thereabouts, F. W. Egan revised the Tertiary lavas and divided them into Upper and Lower 'sheets,' separated by a bed of 'Iron Ore, Bole and Lithomarge'. A revised map was published in 1883. About 1894 Egan returned to the area and revised the 'Silurian' outcrop, dividing it into Upper and Lower groups and a second revision of the map was published in 1898.

About the turn of the century the inadequate drift mapping of the country was recognised and the areas around Dublin, Belfast, Cork, Limerick and Londonderry were remapped for new colour-printed 'Drift' editions. The Belfast Sheet did not coincide with the old one-inch sheet boundaries and only the County Down part of Sheet 29, as far east as Bangor, was revised by A. McHenry and included in the new map which appeared in 1904. A descriptive memoir appeared in the same year.

The revision of Sheet 29 started in 1958 and was completed in 1960 under the supervision of J. A. Robbie, District Geologist. A. E. Griffith and P. I. Manning were the surveyors most involved while H. E. Wilson mapped a small area in the north-west of the sheet and E. J. Cobbing small areas in Antrim and Down. A revised map, in solid and drift editions, was published in 1968.

This memoir, delayed by staff changes, was largely written by A. E. Griffith and H. E. Wilson.

P. A. Sabine and R. R. Harding have contributed sections on the intrusive and extrusive igneous rocks and J. R. Hawkes, E. J. Cobbing and R. A. Old have also worked on the petrography of the igneous rocks while X-ray determinations were provided by B. R. Young and R. J. Merriman. Thanks are due to Dr S. O. Agrell for generously lending material from Carneal. Ordovician and Silurian sandstones were examined by R. W. Sanderson. The Ordovician and some of the Silurian graptolites were identified by Dr I. Strachan; A. W. A. Rushton identified the remainder of the Silurian graptolites and prepared the palaeontological account of the Lower Palaeozoic rocks, in which he was advised and assisted by Dr P. Toghill. Palaeontological determinations of the Carboniferous were by M. Mitchell, Dr J. E. Robinson and F. W. Anderson; the Permian fossils by J. Pattison; the Triassic and Jurassic by H. C. Ivimey-Cook; the Cretaceous by R. V. Melville and C. J. Wood and Recent material by C. J. Wood. The fossil collecting was mainly by R. Carnaghan and T. P. Fletcher.

The stratigraphy of the Carboniferous outcrop was revised by A. Brandon; G. Warrington revised the Triassic and Permian nomenclature; T. P. Fletcher and C. J. Wood examined the Cretaceous and wrote the relevant chapter.

The account of the geophysics of the area was prepared by J. R. P. Bennett. Mr A. Woodrow contributed a historical note on the Conlig lead mines.

Thanks are due to Imperial Chemical Industries and Irish Salt Mining and Exploration Co Ltd for access to records and to workings in the salt field and to numerous quarry owners for access to their workings in County Down.

The memoir has been edited by H. E. Wilson and A. E. Griffith with assistance from R. S. M. Young.

G. M. BROWN
Director of Geological Survey
in Northern Ireland

Institute of Geological Sciences
Exhibition Road
London SW7 2DE
25 July, 1982

PREFACES TO THE FIRST EDITIONS

The descriptions of the area covered by Sheet 29 were originally contained in two separate memoirs, each of which covered an area more extensive than the present sheet.

Memoir to accompany Sheets 37, 38 and part of 29

The northern portion of the district here described, contained in Sheet 29, lying along the shore of Belfast Lough, was surveyed by Mr Du Noyer, under the direction of Professor Jukes, and previous to publication was partly revised by myself in company with Mr Warren, in the spring of 1869. The southern portion of the district, comprising the whole of Sheet 37, was partly surveyed by Mr Du Noyer, but principally by Messrs Warren and W. B. Leonard, and inspected by Mr Jukes. As both Messrs Jukes and Du Noyer were removed by death before they had an opportunity of drawing up the usual descriptive memoir, its preparation has devolved on the authors, who have availed themselves of the information obtained by previous observers, both those connected with the Geological Survey or otherwise.

EDWARD HULL
Director of the Geological Survey of Ireland
Dublin
21 April 1871

Memoir to accompany Sheets 21, 28 and 29

The southern portions of the district here described, and parts of the neighbourhood of Antrim, were surveyed by the late Mr Du Noyer during the years 1867–8, but, owing to his lamented death, were left unfinished. In 1872, however, the survey of the Antrim district was resumed by Mr Duffin, who completed the mapping of the Antrim sheet (No. 28), and would have drawn up the Explanatory Memoir, which it was intended should accompany it, but that he was appointed County Surveyor to the western division of Limerick, and was consequently unable to devote the necessary time to matters belonging to the Geological Survey. Under these circumstances, I found it necessary to undertake the preparation of the Memoir myself, which I have endeavoured to do, partly from my own knowledge of the district, and partly from the notes and memoranda left by Messrs Du Noyer and Duffin.

Mr Bailey has supplied the palaeontological portion, and I have availed myself, to some extent, of the recorded observations of geologists who have written upon this district.

Owing to the circumstances here stated, the Memoir is not as full and complete as it would have been if drawn up by the surveyors actually employed in the survey of the district; but it is hoped it will be found sufficient, when combined with the geological maps, to give a general idea of the structure of the country embraced within the limits of observation.

EDWARD HULL
Geological Survey Office
Dublin
20 January 1876

SIX-INCH MAPS

NOTE

The six-inch to one mile field maps included, wholly or in part, in one-inch Sheet 29 are listed below, with the initials of the Surveyors (E. J. Cobbing, A. E. Griffith, P. I. Manning, H. E. Wilson) and the date of resurvey.

The maps are not published but manuscript copies are available for consultation in the office of the Geological Survey where photocopies can be purchased.

Numbers set in round brackets and preceded by the letters NI refer to rock specimens in the collection of the Northern Ireland Geological Survey; those in square brackets refer to photographs in the collection.

ANTRIM

46 NW SW	H.E.W.	1959
NE	H.E.W., E. J.C.	1959
SE	P.I.M.	1959
47	P.I.M.	1959
52 NW	H.E.W.	1959
NE	P.I.M.	1960
SW SE	E.J.C.	1960
53	P.I.M.	1959

DOWN

1 NW SW	E.J.C.	1960
NE SE	A.E.G.	1959–60
2 NW SW	A.E.G.	1959
NE SE		
3 SW	A.E.G.	1959
5 NW NE	A.E.G.	1960
6 NW NE	A.E.G.	1958–9
7 NW	A.E.G.	1959

All grid references refer to the Irish Grid square J

CHAPTER 1

Introduction

AREA AND LOCATION

The district described in this memoir and covered by the Carrickfergus (29) Sheet of the Geological Survey of Northern Ireland one-inch to one-mile map series lies on both sides of Belfast Lough and includes some 96 km^2 of south-east County Antrim and 88 km^2 of County Down and the Copeland Islands. The City of Belfast, at the head of Belfast Lough, lies just beyond the limits of the Sheet.

PHYSIOGRAPHY

The area is divided into two regions, of very different geomorphological character, by Belfast Lough. North of the Lough the south-east edge of the Antrim plateau looms over a coastal plain, carved from the soft Triassic mudstones, while to the south the glacially moulded Lower Palaeozoic greywackes and shales give a less dramatic rolling countryside, which rises to the Holywood Hills in the south-west (Figure 1). These hills back an extensively industrialised, flat coastal plain and reclaimed land, between Belfast and Holywood, which is underlain by Carboniferous and Mesozoic beds. To the east of Holywood, the coastline is more varied with low cliffs and wave-cut platforms, cut in Ordovician and Silurian grits and shales, separated by small sandy bays. On the County Antrim shore, the area west of Whitehead is low-lying but to the north-east of the town, rugged basalt and chalk cliffs occur, heralding the black and white cliff-line characteristic of the coast further to the north.

In County Antrim the topography is strikingly controlled by the geology. The Tertiary basalt lavas, which cap the high ground in the north-west of the area, overlie the hard Ulster White Limestone (Chalk) of Cretaceous age. This, in turn, rests on a variable succession of Cretaceous sandstones, Jurassic (Lias) mudstones and Triassic mudstones which are intermittently exposed along the base of the scarp and on the flat ground, between the scarp and the coast.

The scarp reaches its maximum height of 275 m at Knockagh. To the east, it is dissected by a series of steep-sided valleys eroded along N–S-trending fault lines. Spurs and outliers of the lavas and chalk, which are more resistant to erosion than the older rocks, form high ground which reaches the coast at White Head, Black Head and Island Magee.

The soft nature of the Jurassic beds, and the spring line formed at the contact between impermeable mudstones and the Cretaceous rocks, give rise to extensive landslipping along the foot of the scarp below Knockagh.

In contrast to County Antrim, the area of County Down within the sheet does not possess dramatic scenic features. The east–west trending Holywood Hills reach a height of 219 m at Cairngaver but the lower ground, about 40 m

above sea level, is covered by boulder clay which is normally distributed in subdued drumlins giving the countryside a pleasant, rolling, open aspect.

Within the area of the sheet there are no large rivers. In County Antrim, drainage is by a number of small streams flowing southwards and eastwards to the sea. The largest, the Woodburn River, has been impounded in its upper reaches to form a series of reservoirs supplying water, mainly to Belfast. The Copeland Water, together with the waters of Lough Mourne, have been similarly impounded. In County Down, a series of short northward flowing streams drain the Holywood Hills and, as in County Antrim, have been impounded. The stream which flows through Holywood has two small dams on it south-east of the town, while further east in Ballymenagh, about 5 km from the town, another unnamed stream has been impounded. In the open valley east of Craigantlet (the spelling Craigogantlet used on the Ordnance Survey maps is unknown elsewhere) two reservoirs collect water from several small streams to supply Bangor.

In the area east of the Holywood Hills the streams flow through the low-lying drumlin country, either northwards to Belfast Lough, or eastwards to the Irish Sea, draining *en route* several large inter-drumlin bogs and alluvial flats. One of these streams, which flows into the sea at Orlock Bridge, has been impounded at Portavo to provide water for Donaghadee.

GEOLOGICAL SEQUENCE

The oldest rocks exposed in the area are the Ordovician greywackes and shales (Figure 2) which outcrop along the south shore of Belfast Lough, in a belt up to 5.5 km wide, extending from Mertoun Hall, near the south-west margin of the sheet to Orlock Point. Similar rocks also occur on the Copeland Islands and in small inliers at Donaghadee and Portavo Point. Silurian rocks, lithologically similar to the Ordovician sediments, underlie the remainder of the County Down area to the south of the Ordovician outcrop.

Towards the end of Silurian times, deformation of the Lower Palaeozoic sediments was initiated and the ensuing polyphase Caledonian orogeny produced intense and varied styles of folding and major faulting. Subsequently, prolonged erosion took place. No Devonian beds are exposed but the existence of rocks of this age, at depth, in the area north of Belfast Lough is probable.

In Carboniferous times, red sandstones were deposited in shallow water and now rest unconformably on the Lower Palaeozoic rocks at Cultra, near Holywood. These sandy beds are succeeded by grey shales and micritic limestones, the latter containing a fauna which reflects a change from continental to marine conditions. Resting on these beds,

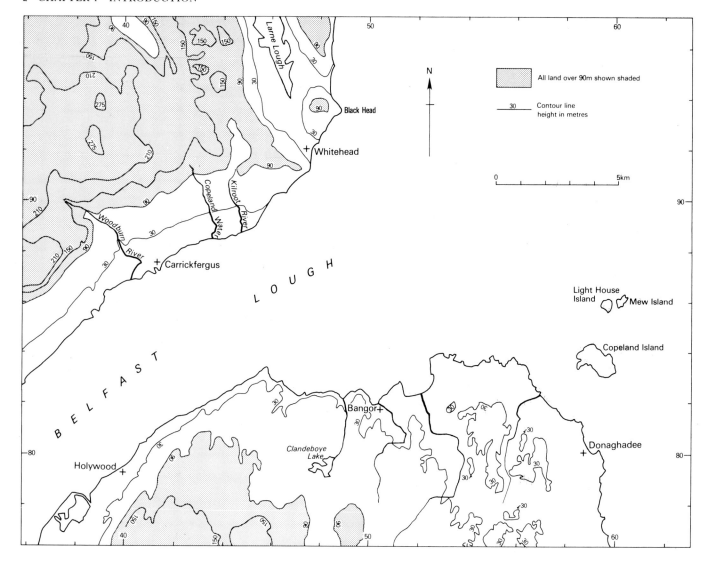

Figure 1 Outline of the main physical features of the district

with a small angular discordance, which, however, represents a considerable time interval, is about 1 m of Basal Permian Brockram and some 10 m of Magnesian Limestone. The latter is considered to have accumulated in an embayment of the Bakevellia Sea. No evaporites are recorded at outcrop in the succeeding Permian Upper Marls, though anhydrite has been proved in the Avoniel Borehole in east Belfast (Sheet 36).

The Permian strata at Holywood are overlain unconformably by Triassic red-beds (Sherwood Sandstone and Mercia Mudstone Groups) but the best development of these is to the north of Belfast Lough. Thick and extensive beds of salt occur to the north and north-east of Carrickfergus, in the middle part of the Mercia Mudstone Group, while small amounts of anhydrite and gypsum are present in all, except the uppermost, parts of the sequence.

Near the end of the Triassic period the environment of deposition of the Mercia Mudstone Group changed and the Collin Glen Formation foreshadows the marine environments in which the Penarth Group and Lower Lias beds

accumulated. These dark coloured shales and mudstones with thin ribs of limestone are only rarely exposed along the foot of the Antrim escarpment and are probably discontinuous due to pre-Cretaceous earth movements and to erosion.

The Lias is overlain by the glauconitic Hibernian Greensands Formation of the Cretaceous, which is also only intermittently exposed and is, in turn, overlain by the Ulster White Limestone Formation. There are non-sequences in the Hibernian Greensands Formation, and pronounced facies variation, indicative of deposition on an unstable floor. Although the succeeding Ulster White Limestone must have been deposited in a sea with virtually no detritus carried into it, the unstable conditions continued, as indicated by minor non-sequences in the chalk.

The Ulster White Limestone is best exposed at White Head and Island Magee, but it also occurs in isolated exposures in the escarpment below the basalts and in valleys.

Late in the Cretaceous, or early in the Tertiary, some crustal warping occurred; the chalk was extensively eroded and a terrestrial deposit of red clay with flints was formed on the weathered limestone surface.

In Eocene times, County Antrim was the centre of a volcanic episode which gave rise to a thick sequence of lavas, some pyroclastic eruptions, and the intrusion of a large number of sills, dykes and volcanic necks. In this area only the Lower Basalt lavas remain, the upper formations having been removed by later Tertiary erosion. There are a number of minor intrusions and at Carneal, a volcanic neck. Large-scale deformation, including faulting on a predominantly NNW-trend, occurred after the extrusion of the Antrim Lava Group. Some of this faulting may follow pre-existing fractures.

During the Pleistocene period, the topography was modified by glacial erosion and a variable cover of boulder clay, in places moulded with drumlins, was laid down over much of the area. The thinnest deposits are on the high ground. Fine-grained red sands, the probable equivalent of the Malone Sands of south Belfast, occur on the coastal plain south-west of Holywood. Small pockets of laminated red clay, which may mark the sites of temporary ice-ponded or sub-glacial lakes, occur in depressions on the north side of the Craigantlet Hills and on the southern slopes of the Antrim escarpment.

In post-Glacial times, fluctuating land and sea levels produced raised beaches and rock platforms which are particularly well developed on the County Down side of Belfast Lough. Estuarine clays underlie the flat land at the head of Belfast and Larne Loughs. Deposits of peat, now mostly cut-out, and alluvium occur in the inter-drumlin hollows of east Down.

PREVIOUS RESEARCH

For obvious geographical reasons, early accounts of the geology of this district were confined to one side or the other of Belfast Lough, and even the first edition of the geological map (1869) was described in two memoirs—one linked to adjoining areas in County Down (Hull, Warren and Leonard, 1871) and the other (Hull, 1876) to adjacent sheets in County Antrim.

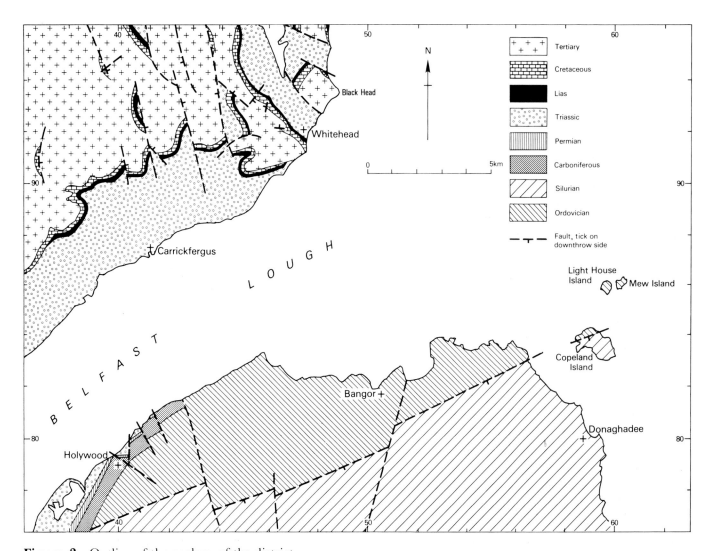

Figure 2 Outline of the geology of the district

The earliest accounts of the country on the north side of the Lough are confined to fairly brief mentions by Bryce (1837) but interest quickened with the discovery of salt in the Carrickfergus area and Doyle (1853) described the Triassic successions. Tate (1864, 1865) gives the first detailed descriptions of the Jurassic and Cretaceous rocks. A revised edition of the Geological Survey map, included some revision of the Antrim Lava Group, was published in 1883.

The most significant contribution to the Cretaceous was by Hume (1897) and after the turn of the century there were papers by Welch (1902, 1904) and others. Within the last two decades a number of important and detailed papers by Patterson (1951a, b; 1952), Patterson and Swaine (1957), and Walker (1960a, b; 1962a, b) on the Tertiary igneous rocks; Reid (1958, 1962, 1971), Fletcher (1978) and Fletcher and Wood (1978) on the Cretaceous; and Stephens (1957, 1958, 1963) on the post-Glacial have increased our knowledge of the area.

On the County Down side, the Carboniferous and Permian rocks on the foreshore at Cultra were recognised early in the nineteenth century (Griffith, 1837, 1843; Bryce, 1852). Haughton (1852) commented on the Conlig mine, first worked at the end of the eighteenth century. The first Geological Survey publication was the Memoir by Hull and others (1871), which covered much of north Down; revision of the Lower Palaeozoic rocks by Egan was followed by a third edition of the one-inch map in 1898. Interest in the Lower Palaeozoic rocks had been developed in the interim, with Swanston and Lapworth publishing important papers in 1877 and Clark reporting on his work in 1902.

In 1902–4 the area as far east as Bangor was included in a revision of the drift deposits by McHenry, Lamplugh and Wright for the Belfast Drift Map published in 1904.

Of the more recent papers, Sharpe (1970) described pillow-lavas in the Ordovician; Adamson and Wilson (1933) and Turner (1952) have mentioned the Carboniferous rocks; and Pattison (1970) has revised the Permian palaeontology. A number of workers, McMillan (1947), Praeger (1892), MacDonald and MacWilliames (1934, 1938, 1961) have considered the post-Glacial fauna in the estuarine clays and raised beaches, and Stephens (1957, 1963) has written on the evolution of the raised beaches of the area.

A small area in the south-east of the sheet is covered by the Special Engineering Geology Sheet of Belfast (3 inches to one mile) published in 1971.

CHAPTER 2

Ordovician

GENERAL ACCOUNT

Ordovician rocks outcrop in north County Down in a belt up to 5.5 km wide along the southern shore of Belfast Lough from Craigavad to Orlock Bridge and on the Copeland Islands. In addition, there are two small inliers at Kittys Altar, 3 km N of Donaghadee and at Coalpit Bay, 1 km S of the town.

On the south of Orlock Point, the southern limit of the main outcrop is a clearly defined fault-line—the Orlock Bridge Fault—which trends 70° and can be traced as a feature for almost one kilometre south-westwards from the coast. Inland, exposure is sparse and beyond the feature mentioned above there is no readily discernible line continuing the Orlock Bridge Fault, and delineation of the boundary is based on differences in the degree of tectonism in scattered quarries and knolls, combined with a knowledge of the regional fault system which extensively dislocates the boundary.

On Copeland Island, the southern boundary of the Ordovician outcrop is a fault trending 70°—the eastward continuation of the Orlock Bridge Fault but offset slightly to the south. To the north, the outcrop is covered by Belfast Lough but indications are that a cover of younger rocks occurs north of a line running north-east from just off Grey Point and swinging south a few kilometres beyond Mew Island.

PREVIOUS RESEARCH

As early as 1816, Berger and Conybeare (1816, p. 146) described the Lower Palaeozoic rocks of County Down as greywackes and compared them with those of the Southern Uplands of Scotland. Subsequently Bryce (1853, pp. 42–43) recorded in the Lower Palaeozoic rocks '. . . a few imperfect fossils, which seem to make them referable to the Lower Silurian Group, but as yet no definite line has been made out to justify a classification.'

The following year Murchison (1854, p. 166) commented 'It is believed that the Schists of Down are of the same age as the Graptolitic Schists of Wigtown and Galloway.'

On the Geological Survey of Ireland One-inch Sheet published in 1869, the Lower Palaeozoic rocks are all classified as 'Lower Silurian', but in the accompanying memoir, Warren (in Hull and others, 1871, p. 24) noted that '. . . three hundred yards N of the Cow and Calf Rocks we get into a new set of beds, . . .' However, as he had no palaeontological evidence, this boundary was not marked on the One-inch Sheet although there are numerous pencil lines on the original six-inch field sheets, some in the position now taken as the boundary fault.

In 1877, Lapworth, working on extensive collections of graptolites made by Swanston, recognised that both Lower and Upper Silurian rocks—that is, both Ordovician and Silurian—were present. However, on the revised edition of the One-inch Geological Survey Sheet, published in 1883, the Lower Palaeozoic rocks were not subdivided but left as Lower Silurian.

The preparation of a memoir on the Silurian Rocks of Great Britain by the Geological Survey revived interest in the rocks of County Down and in 1893 R. Clark started to collect fossils from the whole Lower Palaeozoic outcrop. In March 1894, B. N. Peach, from the Geological Survey of Scotland, met the Director-General, Sir Archibald Geikie, in Dublin and, after examining these fossils in the Geological Survey of Ireland collections and allocating them to their correct stratigraphical positions, they visited County Down with F. W. Egan and R. Clark '. . . where the successive Silurian zones were studied on the ground' (*Annu. Rep. Geol. Surv. G.B.*, 1895, p. 288). As a result of this visit, the Ordovician–Silurian boundary, or as it was then called, the Lower Silurian–Upper Silurian boundary, was drawn across north Down from a small bay south of Orlock Bridge and about 365 m south of the present position of the boundary at the Orlock Bridge Fault. However, from the numerous pencil lines on the field slips, there was obviously considerable doubt about the precise line to choose and the line now chosen as the boundary at Orlock Bridge was only discarded in favour of the more southerly line after many changes of mind. A revised edition of the one-inch map incorporating the division of the Lower Palaeozoic rocks into Upper and Lower Silurian was published in 1898.

During the resurvey in 1959 and 1960 it was noted that, although it was difficult to categorise the lithological differences to the north and south of Warren's original line, there was, even on a cursory examination of the coastal section, a pronounced difference in the degree of tectonism of the rocks on either side of the Orlock Bridge Fault zone. Consequently, the Orlock Bridge Fault zone and its westerly extension along the Donaghadee reservoir gully was regarded as the northern limit of the Silurian outcrop.

Mapping of the remainder of the sheet then proceeded, with particular emphasis on measuring the minor structures to determine why the rocks to the north of the Orlock Bridge Fault were more shattered and cleaved than the Silurian rocks to the south. Detailed measurements of bedding, cleavage, fold axes, fold axial planes, type and orientation of faults and joints were made. The results are presented in Chapter 12 along with the tectonic history.

LITHOLOGY

The Ordovician rocks consist of alternations of arenaceous and argillaceous sediment, with a few occurrences of spilite, agglomerate and ash. Typically, the greywacke beds are various shades of grey but, due to tectonic alignment of abundant mica flakes, mainly in the argillaceous beds, may be glossy silver.

The arenaceous beds are sub-greywakes normally between 0.3 and 1 m thick though they range from a few millimetres to 4 m. The thicker sub-greywacke beds contain rock fragments up to 6 mm in diameter and are described as grits; medium bedded sandstones are medium to very coarse grained while the thinner beds contain predominantly silty material. The grit and sandstone beds vary considerably in texture. Rip-up clasts of mudstone, occasionally up to 0.3 m long but generally 25 mm to 10 cm long and up to 10 mm thick are often incorporated, particularly in the lowest few centimetres, of these sandy beds. Beds, and more commonly lenticles, of fine pebble conglomerates occur. Lenticular patches of calcareous sand and grit also occur and tend to be eroded more readily than the bulk of the bed, leaving an oval depression floored by a soft crumbly sandstone with a honeycomb-like texture.

Graded bedding is extensively developed throughout. In the thicker beds, a basal coarse quartzose grit may grade either continuously or by repeated grading into a current-bedded siltstone or mudstone top. Numerous bottom structures, including flute and groove-casts, associated with flame structures, occur on the soles of massive greywackes along the coastal section between Grey Point and Craigavad.

In the coastal strip west of Grey Point, the attitude of approximately 100 flute and groove-casts was measured and after correction for plunge of the folds and dip of the strata a more westerly group of measurements indicated currents trending north-north-west and an eastern group currents trending north-east. The few readings indicative of the current directions, found only in the western group, indicate that the currents flowed to the south-south-east.

In the predominantly argillaceous beds there are thin ribs of pale grey, buff-weathering, often current-bedded siltstone ranging from 5 to 150 mm thick. The thinnest ribs are commonly of constant thickness while the thicker siltstones vary, and are often current-bedded above a finely laminated base.

The argillaceous sequences consist of bedded mudstones and shales, often with a small percentage of silty material. They are generally grey, though in the thicker, more compact mudstones the prevalent colour is blue-grey. In several areas, however, grey-green, black and red mudstones and shales are exposed. These argillaceous beds

Plate 1 Volcanic agglomerate. Subrounded pebbles of spilite embedded in tuff. Horse Rock, Helens Bay [NI 779]

occur either as partings, of variable thickness, between arenaceous beds, or as units of considerable thickness with many thin current-bedded siltstone ribs. The latter are usually about 3 to 4.5 m thick but may be as much as 42 m as at Garrahan Isle [546 837] 730 m NE of Groomsport village.

In some cases the thin current-bedded siltstones grade up into the mudstones, as can be seen on the coast 450 m WNW of Carnalea Station.

All exposures of the black mudstones are highly brecciated and pyritiferous to a variable extent, with occasional thin layers of pale grey pyrite. In some places, the black mudstones are cherty and occur in association with red or green-grey mudstones.

Ordovician pillow lavas with associated tuffaceous agglomerate and metadolerite were first described by Sharpe (1970, pp. 299–301) from Horse Rock at the western end of Helen's Bay beach [460 830]. The tuffaceous agglomerate (Plate 1) was described by Hull and others (1871, p. 22) as '. . . a conglomerate formed of rounded fragments of felstone in a pale purple and green arenaceous paste.' The succession, which is overturned, dips steeply

south although the nearest greywacke unit exposed in the brecciated sedimentary succession immediately to the north is right way up.

Brecciated metadolerite also occurs on the northern edge of Irish Hill [451 820] 1 km SW of Helen's Bay village and forms a slight feature some 3 m wide which can be traced intermittently along the strike for 0.8 km. Further to the south-west a borehole at Glencraig Bridge [438 814] passed through 2.13 m of spilitic extrusive rock interbedded with greywacke.

Sharpe concluded that the pillow lavas of Helen's Bay underlie Lower Glenkiln shales (*Nemagraptus gracilis* Zone) and are the stratigraphic equivalents of the cherts and tuffs of Morroch Bay in the Rhinns of Galloway rather than the spilitic horizon of Portslogan described by Kelling (1961, p. 41) which lie between two layers of mudstone containing *gracilis* Zone graptolites.

Although the pillow lavas and most of the sedimentary rocks on Grey Point are inverted, it is not obvious whether the volcanics are overlain or underlain by the Glenkiln Shales as gaps occur in the succession. From the identification of graptolites collected from the mudstones overlying

Plate 2 Asymmetrical anticline in Ordovician strata. Foreshore 370 m N of Craigdarragh [NI 395]

the agglomerate, Dr Toghill concludes that although the fauna contains elements suggestive of the *gracilis* Zone much of it indicates *peltifer* Zone, ie Upper Glenkiln. There is a suggestion that other zones of the Lower and Upper Hartfell are present in the succession towards Grey Point to the north but these have not been unequivocally identified and precise correlation with a specific section is not attempted here.

STRUCTURE

The Lower Palaeozoic rocks are intensely folded and there is evidence of three phases of folding. These will be dealt with more fully in Chapter 12. Isoclinal, concentric and occasionally open folding occur and, even when hinges are not exposed, folds can be demonstrated by rapidly changing younging directions across the strike of the predominantly southwards-dipping beds. The axial planes usually dip steeply to the south but on the Copeland Islands they either dip steeply northwards or are vertical. The majority of the folds are periclinal, but in the shore section north of Carnalea Golf Club they plunge at steep angles. In this case the disposition of the axes is due to the rotation of a fault-bounded block.

Slaty cleavage, associated with the main folds, is developed throughout the succession and, less frequently, there is a fracture cleavage confined to the incompetent beds. A low angle strain-slip cleavage, of very limited distribution, is seen in the coastal section between Orlock Point and Carnalea Burn. Occasionally, as in the cliff sections to the south of Orlock Point, the fracture-cleavage is sigmoidally twisted by later concentric shear on the bedding planes between competent and incompetent strata. With the exception of the cleavage on Copeland Island, both slaty and fracture cleavage dip steeply to the south and are parallel or subparallel to the axial planes of the folds, even though they may diverge markedly in the hinges. On the Copeland Islands, the limbs and axial plane cleavages in the isoclinal folds dip steeply north. Most commonly, the fold axes in the Ordovician outcrop trend north-east, but in a few faulted blocks the folds trend eastwards or even east-south-eastwards. From Sheep Point [548 838], east of Groomsport, to near Redburn [397 773] all the folds trend north-east with the exception of a small block about 1.5 km to the south-west of Bangor. The trend of the fold axes around Orlock Point and on Mew and Light House Islands is also north-east. In the small block south-west of Bangor and in the blocks west of Sandeel Bay [555 836] and north of Port Dandy [586 837] on Copeland Island, the fold axes trend approximately eastwards. Between Orlock Tunnel and the 130° trending faults to the south, and in Carn Point [585 831] on Copeland Island, the fold axes trend east-south-east.

Due to the paucity of distinctive marker horizons there are very few major faults on which the exact nature and amount of movement can be demonstrated. However, many small wrench faults and low angle thrusts dissect the coast where they often weather out as sea gullies.   AEG

PETROLOGY

The arenaceous beds are sub-greywackes consisting of subangular to subrounded quartz grains (0.01 to 1.50 mm) with fragments (0.01 to 6.0 mm) of devitrified acid (rhyolitic) lava, cherty material, metaquartzite, perthite, microcline and sodium-feldspar and occasional pieces of intermediate (andesitic) and basic lava, mica-schist and granophyre, set in a recrystallised matrix of sericite, muscovite, chlorite(?), clay constituents, iron oxides, leucoxene and, in some cases, sulphide. Calcite is present in amounts ranging from under 1 per cent to about 5 per cent in the matrices of most rocks. Accessory constituents sometimes present are schorlite tourmaline, zircon and sphene.

There is some variation in the type of clastic material present in the samples. Overall, there is a reduction in the proportion of lithic fragments, with the rocks generally becoming more quartzose towards the south-east. Particular lithologies characterise the clastic fraction in different localities. West of Smelt Mill Bay, the clasts are dominated by plagioclase, hornblende-andesite and albitised basic rocks, some showing fluxioned feldspar laths. Eastwards, acid plutonic and minor intrusive types, quartz-muscovite and quartz-chlorite-schists and potash feldspars are more in evidence. It is possible that the disappearance of the softer, more basic clasts, is partly a post-depositional effect due to low-grade regional metamorphism. Sedimentary clasts, mudstone, siltstone and chert, are of minor importance and sporadic occurrence.

In addition to this general change in composition fragments of distinctive rock types occur at several localities.

A rare, but significant, rock type is encountered in one specimen of sub-greywacke (NI 2124) from Wilsons Point [495 825]. In this section one grain of quartz-muscovite-glaucophane rock and another of similar mineralogy with the addition of garnet were noted. Clastic grains of glaucophane, more or less altered to calcite and chlorite, are also present in this specimen and in a second (I 1347) from the coast near Rockport [435 821]. The grains show marked pleochroism with X colourless, Y grey-blue, Z blue, and variable 2V. A sensible uniaxial figure was obtained from one grain indicating the variety crossite. Among the minor constituents, muscovite and zircon are almost universally present. Pale green and brown hornblende and colourless clinopyroxene, may be common in the lower Ordovician strata rich in basic igneous clasts, together with garnet and brown spinel. Where acid clasts predominate, golden brown tourmaline, and biotite are also present.

The impression obtained is therefore of immature sediments derived from igneous and metamorphic terrains, showing little evidence of reworked older sediments.

Of special interest is the presence of glaucophane-bearing clasts associated with 'spilitic' fragments, which may reflect the hidden Palaeozoic rocks below southern Antrim. The nearest known source of glaucophane occurs at Knockormal in the Girvan–Ballantrae area of Ayrshire, some 70 km NE from Belfast Lough (Bloxam and Allen, 1959). Here glaucophane-schists are derived from spilitic lavas and tuffs of Middle Arenig age and fragments of the same rock types are found in Caradocian conglomerates at Glen App, and in agglomerates at Pinbain. Bloxam and Allen suggested that the development of glaucophane-schists may have been developed on a greater scale than is at present evident.

Glaucophane-bearing rock and mineral grains associated with a suite of fragments petrographically similar to those recorded here, have recently been described (Sanders and Morris, 1978) from probable mid-Ordovician rocks at the western end of the Longford–Down Lower Palaeozoic inlier. The presence of detritus from a high pressure (blue-schist) metamorphic terrain in Ordovician greywackes from localities in Ayrshire, County Down and County Cavan suggests the presence of an early subduction zone situated to the north of the present lower Palaeozoic outcrop, and now reburied by later sediments and volcanic rocks.

Two specimens (NI 2125, 2137) contain fragments of a quartzite containing vermicular chlorite. This rock is characteristic of a horizon in the pre-Cambrian Uriconian Volcanic Series in the Mona complex, while granophyre is known from Anglesey. Many of the fragments in the sub-greywacke could, therefore, have been derived from a pre-Cambrian ridge lying to the south-east (see also Manning and others, 1970, p. 11).

An interesting feature of the sub-greywackes is that the presence of sericite, chlorite and muscovite is indicative of low-grade regional metamorphism. Quartz grains frequently show signs of minor recrystallisation and, in certain specimens (eg NI 2121, 2125, 2129) the quartz grains are elongated so that the rocks have a marked schistose texture. The sericite appears to be derived from alteration products of feldspar and the muscovite from amalgamation of sericite flakes.

Associated with a relatively high degree of reconstitution of the original components is the development of augen structures centred on the more resistant clasts, and a consequent schistosity. This effect is noticeable in both hand specimen and sections of the Ordovician rocks between Bangor and the Orlock Bridge Fault and is more conspicuous in the coarser rocks, eg NI 2125–6. These clasts have developed an equatorial fringe, in the plane of schistosity, of phyllosilicate flakes.

Metamorphic modification has in no case obliterated the original clastic nature of the rocks. On the other hand, it has progressed sufficiently for the Ordovician outcrop to be divided, on the criteria proposed by Turner (1936) for the Chlorite Zone of regional metamorphism, with variation through his Chlorite 1 and Chlorite 2 subzones being discernible. Dr R. Dearnley (personal communication) has noted similar signs of low-grade regional metamorphism in the Lower Palaeozoic greywackes of the Southern Uplands of Scotland. There the metamorphic rocks are confined to fairly narrow belts of country, one of which runs into the sea on the west coast.

The few igneous rocks recorded have spilitic or possibly keratophyric compositions which are probably due to regional metamorphism rather than to differentiation of alkali-rich basic magmas (see Hughes, 1972).

At Horse Rock, a metadolerite and the spilitic pillow-lava consist of laths of plagioclase, which although largely altered to clay mineral are apparently of low relief, suggesting oligoclase or albite-oligoclase compositions. The original ferromagnesian minerals are entirely replaced by green or colourless chlorite material. Relics of ophitic textures are recognisable in the dolerite and basaltic texture in the finer grained lava. Abundant limonitised opaque oxide (? goethite) replaces crystalline iron ore (NI 2683–4).

The rock from a borehole at Glencraig Bridge, where 2.13 m of basaltic material was found, is also spilitic. The bed seems to be inverted, so that the topmost material was probably at the base of the flow. It is composed of augen-shaped areas of partly sericitised sodic plagioclase (albite-oligoclase) laths in radiate clumps with some chloritic patches. These areas are separated by irregular veins of penninite chlorite, and calcite occurs as scattered patches and as cross-cutting veins (NI 2721).

The central portion is similar to the base save that the plagioclase laths are indistinct and may have developed from the devitrification of glassy material. Sericitic alteration is more pronounced and there are scattered crystals and veinlets of quartz, possibly the result of assimilation of sandy material by the lava (NI 2722).

The uppermost part of the flow (lowest in the borehole) is a highly carbonated, calcite-veined, devitrified, glassy extrusive rock. Devitrification has produced radiating tufts (spherulitic structures) which may be composed of plagioclase. Chlorite is scarce suggesting that the rock may be of a keratophyric composition. Small patches of quartz mosaic occur in the carbonate which again suggests the possibility of contamination of the lava by sedimentary material (NI 2723).

RWS, JRH

PALAEONTOLOGY

The Ordovician graptolitic faunas from County Down are essentially similar to those from the Glenkiln and Hartfell Shales of the Southern Uplands of Scotland. The zonal subdivision of these shales proposed by Lapworth, Elles and Wood is as follows:

Upper Hartfell Shales	*Dicellograptus anceps* *D. complanatus*
Lower Hartfell Shales	*Pleurograptus linearis* *Dicranograptus clingani* *Climacograptus wilsoni*
Glenkiln Shales	*Climacograptus peltifer* *Nemagraptus gracilis*

In 1877, Swanston gave an account of several fossiliferous localities in County Down, recognising the equivalents of the Glenkiln and Lower and Upper Hartfell Shales. Review of Swanston's lists of fossils and of material in the Ulster Museum, which has been re-identified by Dr R. B. Rickards, does not suggest any substantial modification of Swanston's conclusions.

In 1902, Clark gave a summary of several further fossil localities and horizons he had discovered in County Down.

During the present resurvey graptolites were detected at seven localities in rocks mapped as Ordovician. Of these, four yielded no determinable specimens, viz. Orlock Point, Crawfordsburn foreshore, Craigavad railway cutting and the faulted inlier at Kittys Altar among the Silurian grits on Portavo Point. The other three localities, Coalpit Bay,

Plate 3 Worm-casts on bedding plane of Ordovician bed. Craigavad
[NI 394]

Carnalea and the foreshore south of Grey Point (Ballygrot), yielded numerous graptolites which are generally in a poor state of preservation but indicate the presence of many of the Scottish zones, though, owing to the intensity of the prevailing folding and faulting, no stratigraphical succession of zones could be made out. Representatives of both Glenkiln and Hartfell Shales were detected at two of the localities, Grey Point and Coalpit Bay. The zones present are as follows:

N. gracilis Zone The collection recorded by Swanston (1877, p. 114) from Ballygrot (Grey Point) includes material suggestive of the *gracilis* Zone, eg *Nemagraptus spp.*, *Dicellograptus divaricatus* and *Didymograptus superstes*; but much of the fauna, which includes also *Dicranograptus nicholsoni* and *Climacograptus peltifer* indicates rather the *peltifer* Zone. Towards the western extremity of the map, at Holywood Glen, Clark (1902a, p. 498) collected a fauna including *Nemagraptus spp.*, *Dicellograptus sextans*, *D. intortus*, *D. moffatensis*, *Dicranograptus ramosus* and *Climacograptus bicornis*, which likewise indicates the *gracilis* and/or *peltifer* Zones as does the fauna from locality 5 (Figure 3) at Coalpit Bay.

C. peltifer Zone Besides Grey Point, assemblages tentatively assigned to the *peltifer* Zone were collected from the Ordovician inlier in Coalpit Bay. Swanston (1877, p. 111, para. 6) refers to a collection from the middle of the inlier which includes *Dicranograptus minimus* and *C. peltifer*. The Survey's collections include *Corynoides curtus*, *Lasiograptus costatus*, *Orthograptus* cf. *apiculatus* and *O.* cf. *intermedius* and were made just north of the dyke which marks the southern limit of the inlier (locality 4, Figure 3).

C. wilsoni Zone This also is doubtfully present just north of the dyke in Coalpit Bay where *Climacograptus brevis*, *Dicellograptus sp.* and *Orthograptus* cf. *intermedius* were collected (locality 4, Figure 3).

D. clingani and *P. linearis* zones Assemblages which may be assignable to one or other of these zones were collected by Swanston (1877, p. 113) from Orlock Point and the shore at Carnalea, at which localities *Leptograptus flaccidus* and *Orthograptus truncatus* were recorded. The Survey's collections from Carnalea, though differing somewhat from Swanston's in the species identified, suggest similar horizons.

D. complanatus Zone The *Dicellograptus forchammeri* recorded by Swanston (1877, p. 111) from a block of black shales in the Ordovician inlier at Coalpit Bay is referable (Lapworth, 1880) to *D. complanatus* and represents the Upper Hartfell '*D. complanatus* Band'. The same species is identified, though with reserve, from the foreshore east of Grey Point.

The *D. anceps* Zone is recorded only from Coalpit Bay where Clark (1902a, p. 499) collected the zonal fossil.   AWAR

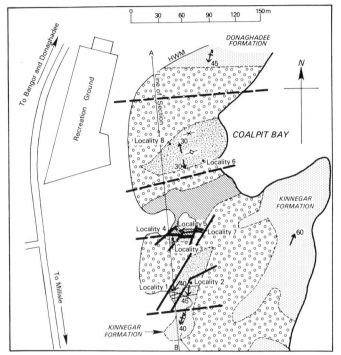

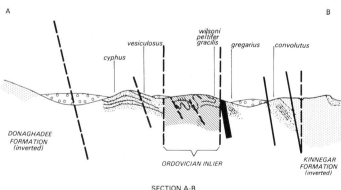

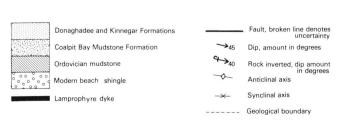

Figure 3 Coalpit Bay. Section and map

STRATIGRAPHY

Any attempt at assembling a stratigraphy for the area is fraught with difficulties. Every exposure of graptolitic mudstone is intensely faulted so that nowhere is it possible to establish a succession and the best that has been possible is to identify, from the graptolite faunas, the occurrence of certain Glenkiln and Hartfell zones.

Similarly, although the graptolitic mudstones occur adjacent to greywacke and volcanic rocks, faulting or non-exposure at the line of contact prevents precise relationships from being determined. Consequently, the following synthesis contains many imponderables and is tentative in the extreme.

The oldest rocks exposed in the Ordovician outcrop may be the association of spilite, agglomerate, metadolerite, red mudstone and chert which occurs at Helen's Bay. It has been postulated (Sharp, 1970, p. 301) that the volcanic rocks occur immediately below the adjacent *gracilis* Zone mudstones on the grounds that the volcanic succession is inverted and that consequently the graptolitic mudstones to the north were originally deposited on top of the volcanics. However, the relationship of these rocks to one another is far from clear. Although the graptolitic beds dip steeply south-east and are, in general, inverted, the greywacke unit closest to the agglomerate is right way up. Furthermore, there is ample evidence of extensive strike faulting so that the volcanic rocks could be within the *gracilis* Zone or older or younger.

Similar problems of succession occur in relating the Carnalea and Orlock graptolitic mudstones to the adjacent greywackes but it is postulated that at least 90 m of greywacke occurs between the *clingani* and *linearis* mudstones at Carnalea and the *peltifer* Zone material at Grey Point. In fact, the mudstones at Grey Point occur associated with greywacke grits which are probably intercalated in the mudstone sequence.

At Carnalea grey-green barren mudstones occur in association with the black graptolitic mudstones but due to the extent of faulting it is not possible to relate them stratigraphically to the fossiliferous mudstones. At Coalpit Bay similar green-grey, barren mudstones dip, in general, steeply south and contain tectonic lenses of black mudstone from the *peltifer* (or *gracilis*), *wilsoni* and *complanatus* zones. In Coalpit Bay the whole Ordovician inlier is extensively brecciated and it is not possible to postulate a succession with any degree of assurance but if the succession across the strike reflects the sequence of deposition then the Coalpit Bay beds are inverted. In the barren mudstones several beds of greywacke grit occur indicating that even this area was subject to turbidity currents carrying in arenaceous sediment.

In summary, the environment of deposition of the Ordovician rocks of County Down is analogous to that prevailing in the area of the Southern Uplands of Scotland in Upper Ordovician times. In the Helen's Bay area, between Horse Rock and Grey Point, the Glenkiln and Hartfell zones of *gracilis* and *peltifer* and possibly *complanatus* are recognised, interbedded in greywacke grits, while at Carnalea, *clingani* and *linearis* Zones are recorded and are probably separated from the beds at Grey Point by at least

90 m of greywacke—a situation similar to that in the northern belt of the Southern Uplands of Scotland.

At Coalpit Bay, even allowing for faulting and tectonic reduction in the thickness of the succession, greywacke units are very thin and the succession, extending from *gracilis* to *anceps*, is probably represented by a thin argillaceous sequence of barren grey-green mudstone and black graptolitic mudstone with only thin intercalations of greywacke indicating that this area, while normally subject to slow gradual sedimentation, was not immune to deposition from turbidity currents.   A E G

Plate 4   Crumpled Ordovician strata. Headland south of Coalpit Bay   [NI 416]

CHAPTER 3

Silurian

GENERAL ACCOUNT

Intensely folded rocks (see Chapter 12) of Silurian age underlie some 57 km² of north County Down and outcrop from the Orlock Bridge Fault to the southern margin of the Sheet. In addition, the Lower Palaeozoic rocks on Copeland Island are mostly Silurian.

The succession is well exposed in the almost continuous shore section from Orlock Bridge [564 829] to Kinnegar Rocks [597 774] and has been divided into four formations; two with abundant arenaceous greywackes, one of graptolitic mudstone and one in which greywacke is interbedded with greenish mudstones. These formations are:

Kinnegar Formation
Coalpit Bay Mudstone Formation
Donaghadee Formation
Portavo Formation

The stratigraphical relationship of the Coalpit Bay Mudstone to the other formations is illustrated in Figure 4 and discussed more fully later (pp. 20–22).

Inland, the Silurian rocks are poorly exposed but it can be shown that the outcrop is divided, by faults, into blocks in each of which the trend of the fold axes, axial plane cleavages and a–c joints are constant.

LITHOLOGY

Portavo Formation

The Portavo Formation probably contains the oldest Silurian greywackes seen within the area of the One-inch Sheet and is best exposed on the coastal rock platform between Orlock Bridge [564 829] and the sandy inlet north of the rocky promontory at Ardgeehan [577 817] 1 km to the south (Figure 5). It also outcrops on Copeland Island, in almost continuous coastal sections, and in the disused slate quarry in Ballygrainey townland [529 785]. In the Ballykeel [41 78] region, south of Holywood, the Portavo Formation is not recognised and may be cut out by faulting which brings the Ordovician and Silurian rocks into juxtaposition, or by facies change.

Typically, the formation consists of two alternating lithological assemblages (Plate 5) one ranging from 15 to 40 m in thickness in which massive arenaceous greywackes are interbedded with thinner shales; the other from 15 to 30 m thick and predominantly of mudstones with thin ribs of siltstone which are locally calcareous or dolomitic and hematite-bearing. The base of the group is not seen as it is faulted against Ordovician rocks by the Orlock Bridge Fault and the top is taken at the change to dominant greywacke grits with thin shales which occurs at Ardgeehan [577 817]. This change is also marked by the disappearance of the thick argillaceous layers characteristic of the group and was also noted by J. L. Warren in the original Geological Survey Memoir of the area published in 1871.

Some of the oldest beds occur in the core of an anticline 200 m SE of Orlock Bridge and are characteristic of the arenaceous greywacke assemblage. They consist of massive grey grits and sandstones with layers of white quartzite pebbles and commonly grade up into siltstone, and are interbedded with thin grey micaceous fissile shales. In the massive beds there are numerous blue-grey, oval, calcareous nodules (Plate 6) with long axes aligned parallel to the main bedding planes. In some cases relict bedding can be traced continuously from a massive bed in a nodule, and occasionally a fragment of another rock-type is found at the core. Thus, these nodules are diagenetic features.

The typical massive 'greywacke assemblage' previously described passes up into mudstones, through a series of thin strongly current-bedded red sandstones (Plate 7) which rest on massive grey grits containing red sandstone nodules.

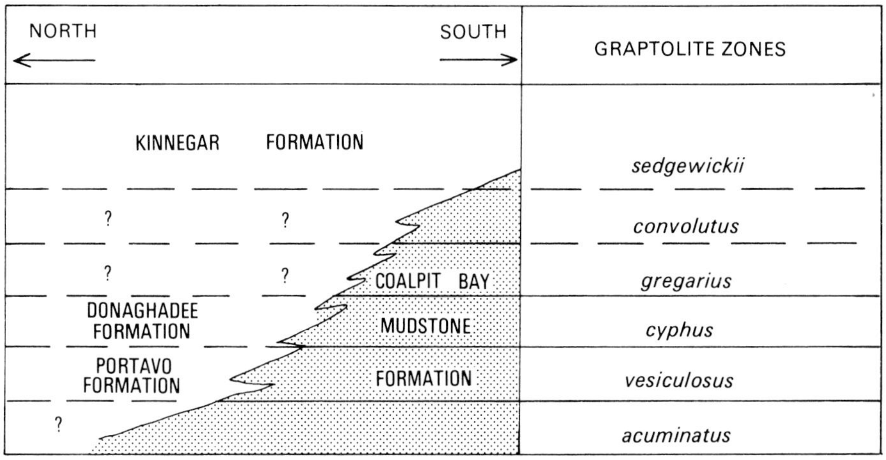

Figure 4 Relationship of the graptolitic mudstones at Coalpit Bay to the greywacke successions

NORTH ←	SOUTH →		GRAPTOLITE ZONES
KINNEGAR	FORMATION		sedgewickii
?	?		convolutus
?	?	COALPIT BAY	gregarius
DONAGHADEE FORMATION		MUDSTONE	cyphus
PORTAVO FORMATION		FORMATION	vesiculosus
?			acuminatus

Plate 5 Portavo Formation. Mudstone with bands of sandstone and siltstone. Cow and Calf headland   [NI 412]

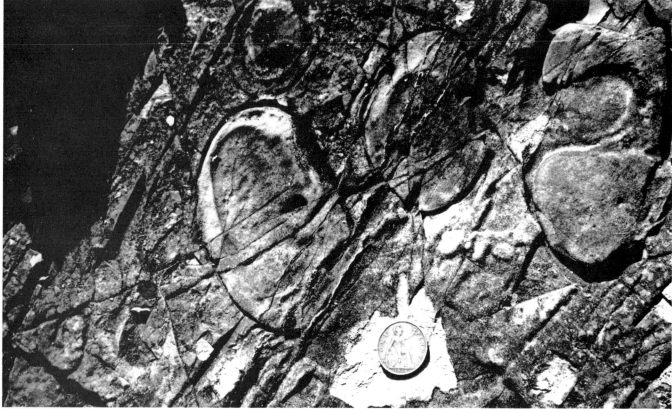

Plate 6 Red concretionary sandy nodules in massive grey sandstone. Foreshore 200 m ESE of Orlock Bridge   [NI 411]

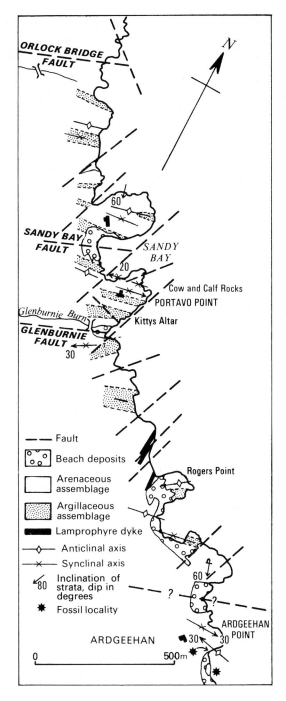

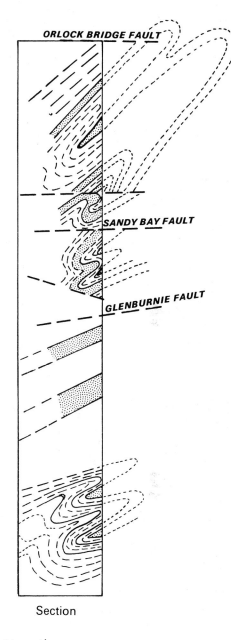

Section

Figure 5 Plan of the foreshore between Orlock Bridge and Ardgeehan Point and stylised section along the coast from the vicinity of Ardgeehan Point to the Orlock Bridge Fault

The 'mudstone assemblage' consists of variably coloured green to blue-grey ribbon-banded mudstones with thin ribs of pale grey to buff-weathering current-bedded siltstone. The thin siltstone ribs vary from 1 mm to 5 mm but are most commonly about 2.5 cm thick.

Flute and groove-casts are well developed on the soles of some greywackes and indicate that the turbidity currents flowed along a north-east to south-west trending axis. From the few examples, in which the direction of flow is determinable, the current direction was towards the south-west.

So far graptolites have only been collected from the Portavo Beds at two localities; on Copeland Island and on the headland north of Sandy Bay. The Copeland Island fauna is sparse but probably indicates the Lower Llandovery

zone of *Monograptus cyphus*.

The beds dip at about 45° to 80° towards 175° and are isoclinally folded with axes essentially horizontal.

There is extensive faulting along the coast section and often the beds close to faults are hematite stained.

Donaghadee Formation

The Donaghadee Formation immediately overlies the Portavo Formation and is well exposed in the almost continuous coast section from Ardgeehan Point to the north shore of Coalpit Bay [594 791]—a distance of about 3 km. In the inland area to the south of the Orlock Bridge Fault, all exposures from Ballykeel [41 78] eastwards to the coast,

Plate 7 Inverted current-bedding in red and green shales, 200 m ESE of
Orlock Bridge [NI 410]

excluding those already attributed to the Portavo Formation, belong to the Donaghadee Formation.

The base of the formation is taken at the change from the
alternating arenaceous and argillaceous assemblages of the
Portavo Formation to the massive sandstone with thin shale
lithology of the overlying beds. Although this change in
lithology is gradual, it is accentuated by strike faulting along
the sandy inlet north of the small rocky promontory at
Ardgeehan and the base of the formation has been taken at
this point. The top of the Donaghadee Formation is not
exposed and the southern limit to the outcrop is taken at the
fault [594 791] cutting off the graptolitic mudstones in the
northern part of Coalpit Bay.

Typically, the formation consists of massive grey and
grey-blue greywacke sandstones (Plate 8) up to 2 m thick
interbedded with thinner mudstones and shales. The massive sandstones are pebbly in places and commonly grade
up into siltstones which in the top 20 cm or so are sporadically cross-bedded. Occasionally, the siltstone top is
missing and the sandstone of the overlying bed rests directly

on sandstone. The sandstones commonly contain calcareous nodules, some of which have dissolved out giving
the rock a carious aspect; others are harder than the
matrix in which they are embedded and stand out on
weathered surfaces so that the rock resembles a coarse
conglomerate when seen from a distance. The nodules are
up to 30 cm long, oval in shape and elongate parallel to the
main bedding planes and may be restricted to discrete
horizons. The soles of the massive greywacke beds are
occasionally marked by flute and groove-casts but these
features are not nearly so strongly developed as in the
overlying Kinnegar Formation.

The shales, which are interbedded with the sandstones
are grey-blue, grey, green or red in colour and are commonly cleaved almost parallel to the bedding planes as on
Foreland Point [587 814]. Particularly in the lower portion
of the formation, the argillaceous beds occur as relatively
thick units with ribs of buff-weathering siltstone and are
very similar to the argillaceous assemblage described previously from the Portavo Formation.

Plate 8 Donaghadee Sandstone Formation: Massive greywacke and thin shales. Robby's Point [NI 415]

Graptolites have been collected from two horizons near the base of the formation at Ardgeehan and at Myrtle Cottage and also from mudstones at the Swimming Pool north of Donaghadee [587 808]. All the fossils recorded are of Lower Llandovery age, about the zone of *M. cyphus.*

Coalpit Bay Mudstone Formation

The Coalpit Bay Mudstone Formation crops out in Coalpit Bay [595 788] 1.25 km S of Donaghadee. To the north the formation is thrown against overturned beds of the Donaghadee Formation (Figure 3) by an east–west fault, while to the south the mudstones are overlain by massive greywacke grits of the Kinnegar Formation. In the intervening area, some 200 m across strike, the outcrop is interrupted by faults, a dyke and a horst of brecciated Ordovician rocks.

The Coalpit Bay Mudstone Formation consists of fine-grained hard, pyritous and somewhat micaceous silty mudstones and siltstones with interleaved thin claystone seams comparable to those in the Birkhill Shales in southern Scotland. These claystone bands have been identified as ash-fall bentonites by Cameron (1977). Three informal members are recognised within the Formation.

Upper Member
Dark grey-black, pyritous, silty mudstones
with thin pale grey and greenish
grey siltstone laminae and ribs,
with abundant graptolites   3.7 m

Middle Member
Grey siltstones with interleaved green and
purple mudstones, and pale buff bentonites;
unfossiliferous   ... seen 1.5 m   ...   estimated 10.7 m

Lower Member
Dark grey-black graptolitic mudstone with
silty laminae and pale claystones   estimated 30.5 m

The graptolites collected from 12 beds within the Upper Member are all from the *M. convolutus* Zone. Although no higher beds were recorded in the present survey, a specimen of *Monograptus sedgwickii* in the Ulster Museum collection may indicate that the *sedgwickii* Zone is present at the very top of the member.

The contact of the Upper Member with the underlying Middle Member is not exposed but is assumed to be conformable as the dip and strike of the beds is similar. Similarly, the contact between the Middle and Lower Members is not seen and, as both units are only intermittently exposed in the shingle and sand-covered bay below high-water mark, the thicknesses quoted above are necessarily estimates and based on the assumption that the section is not affected by faulting. Swanston (1877, p. 119) recorded the presence of *Orthograptus acuminatus*, *Cystograptus vesiculosus* and *Monograptus gregarius* zones from the beds in the Lower Member.

The presence of possible *acuminatus* (Locality 8), *vesiculosus* (Loc. 6), *cyphus* (Loc. 8) and *gregarius* (Loc. 3) Zones has been confirmed elsewhere in Coalpit Bay. Thus a complete succession of graptolite zones from *acuminatus*(?) to possibly the base of *sedgwickii* is postulated and, as in the Birkhill Shales at the type locality at Moffat in the Southern Uplands, numerous bentonitic claystone seams occur indicating contemporaneous volcanic activity.

Kinnegar Formation

The Kinnegar Formation overlies the Coalpit Bay Mudstone and the base of the unit is taken at the lowest massive grit resting in normal superposition on the mudstones in the south-east corner of Coalpit Bay. The top is not exposed in this area. The formation consists of a succession of alternating greywacke grits, sandstones and mudstones and differs from the Donaghadee Formation in the more massive character and coarseness of the grits and the abundant sedimentary structures (Plates 9 and 10) preserved on the soles of greywacke beds.

Characteristically, the greywackes are very massive, grey grits and sandstones up to 6 m thick with thin layers of darker grey, micaceous shales and mudstones most commonly about 30 cm thick.

The grits contain calcareous nodules, though not so abundantly as in the Donaghadee Formation. These nodules are blue-grey in colour, exceedingly fine grained, highly calcareous and have often formed around a nucleus of foreign material such as a mudstone raft.

The most distinctive feature of the formation, apart from the massiveness of the beds, is the prolific development of sole-markings on the greywacke units. These sole-markings, originating as flute, groove and interference ripple-marks have been subsequently accentuated by differential loading.

Plate 9 Flute-casts on under-surface of massive inverted greywacke. Foreshore 140 m S of Galloways Burn [NI 421]

Plate 10   Groove-casts on sole of massive greywacke. Foreshore 140 m N of Galloways Burn   [NI 419]

Rose diagrams plotted for the trend of sole-markings, indicated that currents trended almost due west for the beds around Galloway's Bridge [593 786] and to just west and east of south in the Kinnegar Rocks exposures.   AEG

PETROLOGY

Sandstones from the Silurian outcrop south of the Orlock Bridge Fault zone are poorly sorted greywackes with angular to subrounded clasts of medium to fine sand-grade material. Pebbles measuring several millimetres across are not uncommon. The fine-grained matrix comprises some 35 to 40 per cent of the rocks.

The rocks have been subjected to low-grade, regional burial, metamorphism which has caused recrystallisation of the matrix and some modification of the forms of the labile clasts. The matrices are now composed of chlorite, a little secondary mica together with variable development of turbid 'leucoxene' and granular sphene and redistributed quartz and carbonate.

Clasts exhibit more or less gradational margins against the matrix. The sand-grade particles are dominated by quartz and alkali-feldspars, the latter often turbid with alteration products and sometimes replaced by secondary mica or calcite. Lithic fragments are abundant. In samples from the Portavo Formation (I 1352–3, NI 1338, 2128–9) and the succeeding Donaghadee Formation (I 1336–7, NI 1353, 2130) they are predominantly of acid-intermediate lavas, among which felsitic, andesitic, micro-diorite and trachytic textural types can be found, with minor metamorphic (mica-schist and quartz-epidote rocks), granophyre, mudstone and crystalline limestone fragments. Granitic rock fragments and minerals (including perthite and microcline) become important in the younger Donaghadee and Kinnegar formations (I 1334–5, NI 2131), although acid-intermediate lavas are still represented.

Accessory detrital minerals in the lower horizons include colourless clinopyroxene, blue-green amphibole, epidote, clinozoisite and muscovite. Tourmaline, biotite, muscovite, zircon and garnet were recorded in rocks from the higher formations.

Although specimens from the Portavo Formation may contain sufficient pyrogenetic material to be regarded as volcanic greywackes, eg I 1337 from Portavo [576 818], true lithic tuffs have not been recognised.

Regional metamorphic effects in the Silurian rocks are much less pronounced than in the Ordovician outcrop and the development of schistose structure has been noted in only a single specimen from Portavo (NI 2129). A specimen of pebbly greywacke from near Echlin Grove [520 793] (I 1353) shows extreme alteration of the matrix to turbid 'leucoxene' in which distinct grains of sphene are recognisable. Prehnite has also developed in this rock in veins, as infilling small, about 0.2 mm diameter vugs rimmed with granular sphene and probably also in the matrix. Although pumpellyite has not been recognised, the presence of prehnite suggests that the rocks of northern County Down have been subjected to prehnite-pumpellyite facies burial metamorphism, in common with rocks of comparable age from County Cavan and the Southern Uplands of Scotland (Oliver, 1978).   RWS

PALAEONTOLOGY

The faunas of the Silurian rocks on the Carrickfergus Sheet are appropriately compared with those of the Birkhill Shales in the Southern Uplands of Scotland. The zonal arrangement given by Toghill (1968b), following that of Lapworth, Elles and Wood, is:

Upper Llandovery (part)	Monograptus turriculatus Rastrites maximus M. sedgwickii	Fronian
Middle Llandovery	M. convolutus M. gregarius	Idwian
Lower Llandovery	M. cyphus Cystograptus vesiculosus Orthograptus acuminatus Glyptograptus persculptus	Rhuddanian

Toghill's arrangement is followed here with the difference that the vesiculosus Zone is less restricted, being extended to include the ranges of Dimorphograptus swanstoni, D. erectus and D. decussatus.

The small fossil assemblages from the Portavo Formation and Donaghadee Formation contain Climacograptus and slender monograptids but lack other elements. Dr I. Strachan regards them as probably of Lower to Middle Llandovery age and more probably cyphus than gregarius Zone.

The faunas of the Coalpit Bay Mudstone are considerably richer; in Coalpit Bay (Figure 3) all the zones from the vesiculosus to convolutus are present and it is possible that parts of the acuminatus and sedgwickii zones may also be

represented. The presence of the acuminatus Zone in the vicinity of Locality 8 is suggested by a slab with Orthograptus cf. acuminatus but the bulk of the collection from the locality is in material of different lithology and indicates the cyphus Zone and contains:

Climacograptus medius, C. cf. miserabilis, C. cf. normalis, Diplograptus sp., cf. Glyptograptus tamariscus, cf. Rhaphidograptus toernquisti, Monograptus cf. atavus, M. aff. cyphus, M. gregarius, M. hipposideros Toghill (1968a), M. cf. incommodus, M. cf. revolutus, M. cf. revolutus austerus, M. cf. sandersoni.

The vesiculosus Zone is present at Locality 6 where the following were collected:

Climacograptus rectangularis, Dimorphograptus confertus, cf. D. decussatus, D. cf. erectus, Monograptus sp.

The gregarius Zone at Locality 3 yields the following forms:

Climacograptus rectangularis, Glyptograptus cf. enodis, G. tamariscus s.l., Orthograptus aff. cyperoides, O. aff. insectiformis, O. mutabilis, Pseudoclimacograptus (Metaclimacograptus) cf. hughesi, Rhaphidograptus toernquisti, Monograptus cf. atavus, M. cf. cyphus, M. cf. difformis, M. gregarius [common], M. incommodus, M. revolutus s.l. [common], M. sandersoni, M. triangulatus s.l.

The presence of triangulate monograptids and the abundance of M. gregarius indicate the gregarius Zone; the presence of slender Monograptus of the atavus and sandersoni kind suggest a low horizon in the Zone.

In the collection of the Ulster Museum there are specimens of Orthograptus insectiformis, Petalograptus ovatoelongatus and Monograptus triangulatus fimbriatus from Coalpit Bay; the exact location from which they were collected is unknown but they suggest a higher horizon in the gregarius Zone than that of our locality 3, and may have come from an exposure, now shingle-covered, immediately to the south (Figure 3).

The M. convolutus Zone is present at Localities 1 and 2, the former yielding large collections at twelve horizons (A–L) in a succession 3.7 m thick (Table 1). No equivalent was found of the Monograptus clingani band such as occurs in the Birkhill Shales, and there are differences in the proportions of species which make up the fauna at Coalpit Bay as compared with Dobb's Linn (Toghill, 1968b, p. 661): thus at Coalpit Bay M. leptotheca (Rickards and Rushton, 1968) dominates horizon H in the middle of the Zone whereas at Dobb's Linn its main occurrence is at the top of the underlying gregarius Zone; M. jaculum is a more important constituent of the upper part of the convolutus Zone at Coalpit Bay than at Dobb's Linn, and M. lobiferus is commoner and ranges higher; Cephalograptus tubulariformis is a commoner fossil than C. cf. cometa in the Survey collections but the latter, which is abundant at Dobb's Linn, is also present in the collections of the Ulster Museum and was recorded from the top of the Coalpit Bay succession by Swanston (1877, p. 110).

In the Ulster Museum is a specimen of Monograptus sedgwickii from Coalpit Bay, the same species also being recorded by Swanston (1877, p. 110) from the top of the Coalpit Bay Mudstones. This species is thought to be diagnostic of the sedgwickii Zone; if so, the zone may be present in a mudstone facies just below the greywacke grits at

Table I Occurrence of graptolites at localities 1 and 2, Coalpit Bay, Donaghadee, Co. Down

	A	B	C	D	E	F	G	H	I	J	K	L	Loc. 2
DIPLOGRAPTIDAE													
Cephalograptus cf. *cometa*	.	.	.	.	E	.	.	.	.	J	.	.	.
C. tubulariformis	cf.	B	.	.	E	.	.	.	.	.	.	.	2
Climacograptus scalaris	cf.	B	.	.	cf.	.	cf.	cf.	.	.	.	.	2
C cf. *miserabilis*	.	.	C	.	.	.	.	.	.	.	.	.	.
C. sp.	.	.	.	.	.	.	.	H	.	.	.	.	.
Glyptograptus cf. *incertus*	.	.	?	.	E	.	.	H	.	.	.	.	.
G. sinuatus	.	.	C	.	E	.	.	cf.	.	.	cf.	.	.
G. tamariscus s.str.	.	.	.	.	E	.	G	H	I	.	.	.	.
G. tamariscus s.l.	.	.	.	D	.	.	.	.	.	.	cf.	.	2
G. sp.	.	.	.	.	E	.	.	H	.	.	.	.	.
Orthograptus bellulus	.	.	.	.	?	.	?	H	.	.	cf.	.	2
O. cf. *cyperoides*	A	.	.	.	.	.	.	.	.	.	.	.	.
Petalograptus folium	.	B	C	.	.	.	.	.	.	.	.	.	.
P. palmeus	.	.	.	.	.	?	cf.	.	.	.	K	.	.
P. palmeus cf. *latus*	.	.	C	.	.	.	.	.	.	.	.	.	.
P. palmeus tenuis	.	.	.	.	.	.	G	H	?	.	cf.	.	.
P. sp.	.	.	.	.	E	.	.	.	.	.	.	L	2
Pseudoclimacograptus (Clinoclimaco-graptus) retroversus	A	.	.	.	.	.	.	H	.	.	.	.	.
P. (Metaclimacogr.) hughesi	.	?	.	.	.	.	.	.	.	.	.	.	.
MONOGRAPTIDAE													
cf. *Diversograptus? capillaris*	.	.	.	.	.	.	.	.	.	J	K	L	2
Monograptus argutus	.	?	.	.	.	.	.	.	.	.	.	.	.
M. clingani	.	.	.	.	E	F	G	H	.	.	.	.	.
M. cf. *communis* s.l.	A	.	.	.	E	.	G	.	.	.	.	.	.
M. concinnus	.	.	.	.	.	.	cf.	?	cf.	cf.	cf.	L	.
M. convolutus	.	.	C	cf.	cf.	.	.	cf.	.	.	.	.	2
M. cf. *crenularis*	.	.	.	.	.	.	G	H	.	.	.	.	2
M. decipiens	.	.	.	?	.	.	.	H	I	J	cf.	cf.	2
M. denticulatus	A	B	.	?	E	cf.	G	.	.	.	.	.	.
M. involutus	A	cf.	.	.	.	.	G	cf.	.	.	.	.	.
M. jaculum	.	.	.	.	E	F	G	H	I	J	K	L	2
M. leptotheca	A	.	.	.	cf.	.	?	H	I	cf.	K	cf.	cf.
M. limatulus	A	.	.	.	E	F	G	H	I	.	K	cf.	2
M. lobiferus	A	B	C	D	E	.	G	H	.	J	K	L	2
M. millepeda	.	cf.	.	.	?	.	.	.	.	.	.	.	.
M. regularis	.	cf.	.	.	cf.	.	G	cf.	cf.	.	K	cf.	cf.
M. triangulatus s.l.	cf.	cf.	C	cf.	cf.	.	.	H	.	J	.	cf.	2
M. undulatus	.	.	.	.	.	cf.	G	.	.	.	.	.	.
M. cf. *urceolus*	.	.	.	.	.	.	.	H	.	.	.	.	.
Rastrites peregrinus	.	.	.	cf.	cf.	.	cf.	H	.	.	.	.	cf.
Rastrites spp. [indet.]	A	B	C	.	.	.	.	.	.	.	.	.	.
	A	B	C	D	E	F	G	H	I	J	K	L	2

the south end of the bay, though the Survey collecting did not confirm this.

On the assumption that the Portavo and the Donaghadee Formations are approximately of *vesiculosus* or *cyphus* Zone age and that the Kinnegar Grit is *sedgwickii* Zone and possibly later, the regional stratigraphical setting of the Silurian greywackes of northern County Down may be compared with that of the similar beds in the Southern Uplands of Scotland.

In the Southern Uplands the greywackes overlie the dark graptolitic Birkhill Shales; but Lapworth observed that in passing north-westwards across the strike the greywackes appear lower and lower in the succession until they replace the Birkhill Shales entirely and rest directly upon Hartfell

Shales (Lapworth, 1878, p. 341). At Dobb's Linn, which is some 24 km SE of the Southern Uplands Fault, the Gala Greywackes rest on part of the *Rastrites maximus* Zone (Toghill, 1968b, p. 665). At Hartfell Spa and the Meggat Valley, both about 6.5 km NW of a strike-line through Dobb's Linn, the greywackes rest on the *gregarius* Zone and beds yield *Monograptus lobiferus* (= *convolutus* Zone?) (Lapworth, 1878, pp. 293, 297). Similarly, in County Down the *maximus* Zone is known only from the southern region, at Tieveshilly at the southern end of Strangford Lough (Clark, 1902a, p. 499), whereas at Coalpit Bay which lies about on the strike of Hartfell Spa, the greywacke grits appear at the base of the *sedgwickii* Zone. North from Coalpit Bay, greywackes of *convolutus* and *gregarius* Zone age might be expected but there is no evidence for these; the fault at the north end of Coalpit Bay brings up the Donaghadee Sandstones which, though undated at this locality, are of *cyphus* Zone age 2.2 km further north. It may be that the fault has cut out the higher greywackes. The approximate theoretical relations of the Coalpit Bay Mudstones and the greywacke groups of County Down can be expressed diagrammatically as in Figure 4.   AWAR

CHAPTER 4

Lower Carboniferous

INTRODUCTION

Lower Carboniferous [Chadian] rocks occur in well exposed shore sections near Cultra, on the south side of Belfast Lough, and in some less perfectly exposed stream sections near Cultra and Holywood. They are thought to lie unconformably on Ordovician rocks and to be similarly overlain by Permian beds. The strata, amounting to about 280 m in the exposed part, have been named the Holywood Group. This consists of a series of red and yellow micaceous sandstones passing up, through a transition, into a complex series of thin-bedded shales, mudstones, siltstones, sandstones and fine-grained micritic, dolomitic limestones (termed cementstones in Scotland). Many thin evaporite replacement beds are present in the upper part of the sequence. The Holywood Group can therefore be divided into two formations, the Craigavad Sandstone Formation and the overlying Ballycultra Formation, each approximately 140 m thick (Figure 6). The formations are named from townlands at Cultra.   A B

HISTORICAL ACCOUNT

The isolated outcrop of Carboniferous beds at Cultra has attracted attention from the earliest days of Irish geology. The beds are closely associated with exposures of Permian strata and consequently the literature on the Permian and Carboniferous at Cultra is confused (see Chapter 5) with, on occasion, Carboniferous strata being assigned to the Permian, and *vice versa*.

The first account of the Carboniferous at Cultra was given by Sir Richard Griffith, (1843, p. 45), in which he stated that, although he had previously considered the strata to be of Permian age, he was now convinced that it was Carboniferous on the basis of the fauna. He described a traverse across the section, though from his description it seems likely that he included some of the Permian strata at the top.

The area was systematically surveyed by G. V. Du Noyer, W. B. Leonard and J. L. Warren of the Geological Survey of Ireland in 1867–69. The Survey recognised that both Carboniferous and Permian strata were present on the foreshore at Cultra and published faunal lists from both systems. The Carboniferous succession was referred to the Lower Carboniferous Shale on the map and, in the memoir, to the Lower Limestone Shale. A detailed description of the coast section from the faulted boundary with the Triassic to the faulted boundary with the Ordovician was given. The succession through the basal Carboniferous sandstones in the Cultra Glen was not, however, described and consequently the nature of the junction between the Carboniferous and the Ordovician rocks was not discussed.

A description of the limestone outlier at Castle Espie just to the south on Sheet 37 (Hull and others, 1871, p. 10) was included in the same memoir and the strikingly different faunal lists from this locality are given.

Shortly after the appearance of the memoir, the identification of Permian fossils from Cultra was questioned by Anderson (1873, p. 45) and Wright (1872, p. 34) who considered that all the strata at Cultra were of Carboniferous age.

This matter was finally cleared up by Lamplugh (1904, p. 17) who convinced Wright that the fossils collected by him were all obtained from the Carboniferous portion of the outcrop, and that his argument against the occurrence of Permian rocks at Cultra was based on a misunderstanding.

Wright (1872, p. 35) recorded *Lithodendron junceum*, Flem sp. which in the 1904 memoir was referred to by Lamplugh as *Lithostrotion junceum*, Flem. This is a curious record, as the facies represented by the sediments at Cultra would not normally be expected to contain corals. From Wright's description of the locality (p. 36), it seems clear that his specimen came from a flaggy limestone crowded with serpulids and Wright himself mentioned the possibility of *Serpula* in this connection. Under these circumstances, it seems best to regard the record of *Lithostrotion junceum* from Cultra with reserve. This fossil was not found during the present resurvey.

The resurvey of the area in 1904 was mainly concerned with the drift deposits and no new knowledge of the Carboniferous succession was gained.

Scales of the fish *Holoptychius* have been recorded by Portlock (1843, p. 46), Anderson (1873, p. 46) and Wright (1872, p. 35). Lamplugh and others (1904, p. 13) referred to it as *Strepsodus (Holoptychius) portlocki*, Ag., and mention that it had been recorded by the Geological Survey in 1871, but it does appear in the faunal lists of the 1871 memoir. *Holoptychius* is a genus normally regarded as being of Devonian age. In Scotland, red beds with *Holoptychius* scales which underlie the cementstones were assigned to the Upper Devonian on this basis. Griffith described it as being found in grey shales associated with a normal Carboniferous fauna. The other records are more ambiguous and describe it in both the red beds and the grey shales.

Davis (1890, pp. 332–334) examined a collection by C. Bulla from the dark shales and recorded *Anodontacanthus attenuatus* and *Coltacaninus sp.*

A description of the petrography of the Carboniferous sediments was given by Adamson and Wilson (1933, p. 186), who demonstrated that the clastic material of the sediments had all been derived from the Ordovician rocks to the south.

A review by Turner (1952, p. 140) expressed the opinion that the Carboniferous strata at Cultra were of *Modiola*-phase type in Lower Limestone Shales of Viséan age. In an unpublished thesis, Freshney (1961) recognised the cementstones (micrites or calcilutites) at Cultra and from ostra-

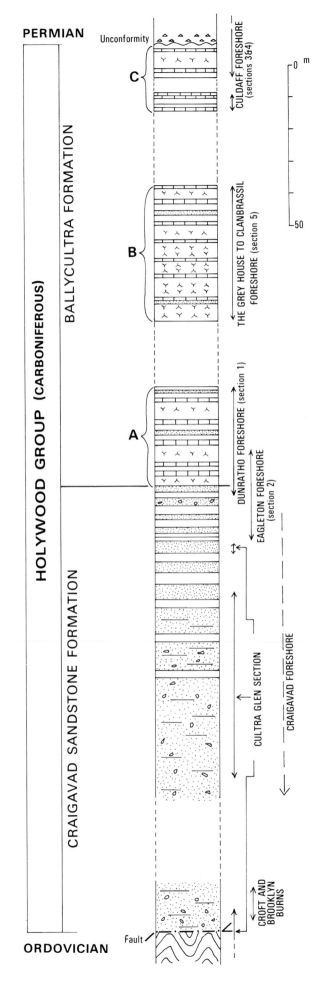

PERMIAN

Unconformity

C

HOLYWOOD GROUP (CARBONIFEROUS)

BALLYCULTRA FORMATION

CULDAFF FORESHORE (sections 3 & 4)

B

THE GREY HOUSE TO CLANBRASSIL FORESHORE (section 5)

A

DUNRATHO FORESHORE (section 1)

EAGLETON FORESHORE (section 2)

CRAIGAVAD SANDSTONE FORMATION

CULTRA GLEN SECTION

CRAIGAVAD FORESHORE

CROFT AND BROOKLYN BURNS

Fault

ORDOVICIAN

0 m

50

Red-stained shales with thin micrites and flaggy bio-clastic limestones; one known evaporite replacement bed; shales contain *Modiolus latus*, ostracods, *Serpula*, *Spirorbis* and fish.

Grey serpulid shales and thin oncolitic micrites; some calcareous sandstones.

Strata, concealed by shingle and till on foreshore, estimated to be approximately 24m thick.

Predominantly dark grey shales and mudstones and greenish grey silty shales and siltstones, reddish only near base; many thin olive-grey compact micrites and calcareous sandstones; fourteen evaporites associated with thin laminated stromatolitic micrites; shales frequently fossiliferous with common *Modiolus*, ostracods, *Spirorbis*, and plant fragments and rare *Camarotoechia* and '*Estheria*'.

Strata partly representative of this level are poorly exposed and faulted on the Grey House foreshore and consist of about 15m of red and grey shales with many sandstones including a 0.47m very massive bed; at least one evaporite.

Predominantly reddish grey, silty, micaceous shales and siltstone becoming increasingly greenish grey upwards; grey eventually predominant. Many thin olive-grey micrites and reddish sandstones. Three evaporites known. *Modiolus*, annelids and ostracods on some levels.

Thin-bedded calcareous, micaceous, ripple-marked, mainly red sandstones and siltstones interbedded with mainly red calcareous, silty, generally unfossiliferous shales; very few micrites; no evaporites.

Red and yellow, medium-grained, frequently pebbly, thick-bedded, micaceous sandstones, becoming thinner-bedded, micaceous sandstones, becoming thinner-bedded upwards where interbedded with red and greenish grey calcareous shaly mudstones; cornstones are frequent but other limestones uncommon.

Massive, red, micaceous, pebbly sandstones with breccia beds.

Figure 6
Carboniferous succession at Cultra

cods in the associated shales gave a Tournaisian age to the succession. He also noted the similarity of these deposits with those in Ayrshire and Northern England, and suggested a correlation.   EJC

GENERAL SUCCESSION AND LITHOLOGY

Craigavad Sandstone Formation

The generalised sequence and relative stratigraphical position of the various sections are summarised in Figure 6 and described in Appendix 1.

The junction between the Carboniferous strata and the steeply dipping Ordovician rocks is not seen, but exposures close to it occur in Cultra Glen. Here, sandstones dipping at 10° are separated from sheared Ordovician rocks by a 2 m gap filled by slumped till. From the disposition of the sandstones it is evident that this gap contains a fault and that the lowest beds of the formation have been faulted out. A faulted junction against Ordovician rocks also occurs on the foreshore, though here the sandstones are higher in the succession than those in the Cultra Glen, and the fault strikes north-westwards rather than north-eastwards. Whilst at the only two places where the junction between the Carboniferous and the Ordovician rocks is approached it is probably faulted, the main contact is almost certainly an unconformity. Accordingly, the junction between the two systems is shown on the published map as a natural one over most of its length, the faulting being indicated only where proved. There is a gradual increase in dip, from 10° at the junction of the Craigavad Sandstone with the Ordovician rocks, to 30° at the passage beds into the Ballycultra Formation.   EJC

The lowest beds are poorly exposed, massive, coarse-grained, friable, red, micaceous, pebbly sandstones with a calcareous breccia bed. Throughout the formation there is a tendency for the flaggy bedded calcareous sandstone units to become progressively thinner upwards. Trains of calcareous nodules (cornstones), about 5 cm thick, are present at some horizons. There are also occasional 15 to 30 cm thick beds of micritic limestone and dolomite and there are also several beds of coarse pebbly sandstone, the pebbles being both of vein-quartz and Lower Palaeozoic greywacke. Extensive ripple-marking is seen in the coastal exposures of the upper beds. Apart from some carbonaceous plant debris in the shales, the formation is generally unfossiliferous, although some of the higher shaly beds on the foreshore contain sparse *Modiolus* and fish debris and a 45 cm shale bed 1.05 m from the top contains abundant ostracods and annelids.

The thin evaporite beds that characterise the overlying formation are absent from the Craigavad Sandstone.

The intense red colouration of the sandstones, the minor green beds, the presence of ripple-marked surfaces, breccia and pebble beds and cornstones, are all characteristic features of near-shore, high energy, oxygenated, shallow water sediments. The complete absence of evaporite beds or stromatolitic micrites suggests that the salinity was not high, while the almost complete absence of fossils, except plants, indicates a restricted or non-marine environment. Therefore, the strata of the Craigavad Sandstone were, probably, laid down in a brackish water environment, hypersalinity being averted by the continual influx of sand-charged fresh water from rivers draining off the Lower Palaeozoic land-mass to the south (Adamson and Wilson, 1933). Towards the close of the period of deposition of the sandstones, conditions approached those which prevailed during the deposition of the Ballycultra Formation.   AB

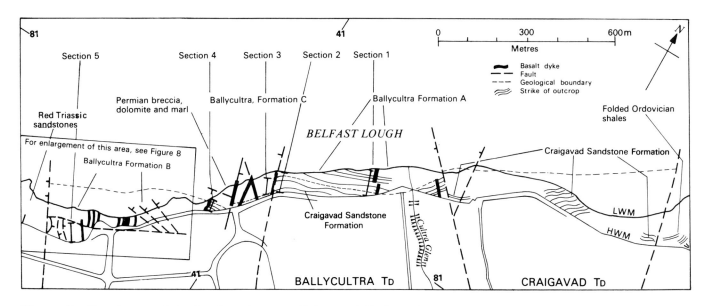

Figure 7 Sketch-map of the foreshore between Cultra and Craigavad, indicating Carboniferous and Permian outcrops and locations of measured sections

Ballycultra Formation

These beds are well exposed in a series of faulted sections on the foreshore at Cultra (Figures 7 and 8) but, as indicated in Figure 6, it has not been found possible to build up a complete succession and there remain two major portions of the formation for which there are no exposed representatives or for which the strata are too disrupted by faulting, to allow a reliable succession to be drawn up. Until these gaps are filled the formation is subdivided into three divisions. These are as follows:

Division C		21 m
Concealed strata	approx.	24 m
Division B		43 m
Concealed and disrupted strata	approx.	20 m
Division A		32 m

Five reasonably complete sections (Appendix 1) were measured by E. J. Cobbing at locations shown on Figure 7. Sections 1 and 2 give overlapping successions through the upper beds of the Craigavad Sandstone and Division A. Sections 3 and 4 provide sections in Division C while Section 5 covers Division B. Subsequently, after the discovery of thin calcitised evaporites in 1973, the sections were re-examined by A. Brandon. In Sections 1, 2 and 4, the original details have been modified only slightly. In 1973 Section 3 was found to be largely covered by drifting shingle. The foreshore around Section 5, however, was found to be so cut up by faults that a complete revision of Division B was undertaken and the original section extended downwards.

The Ballycultra Formation consists of a series of thin-bedded dark grey and greenish grey shales and mudstones, greenish grey siltstones, and micaceous sandstones, thin olive-grey, compact, non-laminated dolomitic micrites or calcilutites (cementstones) and laminated olive-grey dolomitic micrites of (?)algal origin associated with thin, nodular, pink beds of sparry calcite, which replaces gypsum or anhydrite.

Deposition was cyclical and all lithologies are calcareous. Red colouration, due to original depositional factors, which is so characteristic of the Craigavad Sandstone, is common in the lowest part of Division A but upwards greenish grey and dark grey colours become prominent and the red colour is not present above the lowest part of Division B. The red, pink, and orange colours seen towards the top of Division C are due to secondary staining from the overlying Permian.

From a study of Division A, on the foreshore between Sections 1 and 2 and also from a more detailed investigation of Division B, it is apparent that individual lithological bands maintain their character and thickness for considerable distances. Only the sandstone units show significant lateral changes. The evaporite replacement beds and associated laminated micrites, in particular, show characters peculiar to each band and the existence of these marker beds establish the distinctiveness of each division.

Though broadly transitional the lower boundary of the Ballycultra Formation, as exposed on the foreshore around Sections 1 and 2, is clearly demarcated. The junction is taken at the base of a prominent 1 m thick deep greyish red micaceous, calcareous, silty, shale bed.

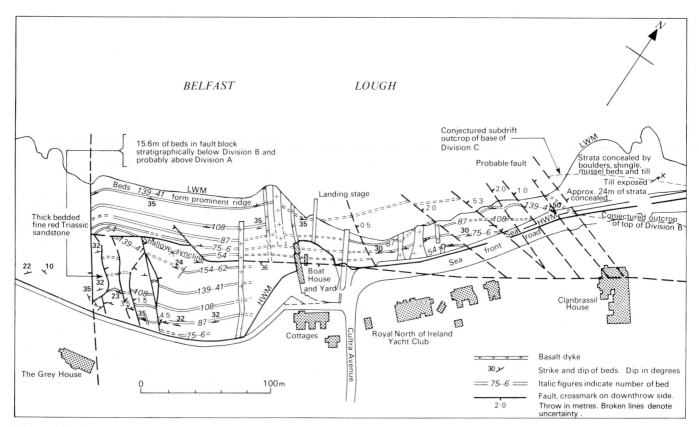

Figure 8 Details of the foreshore exposures at Cultra

The boundary between the Carboniferous and the overlying Permian is well exposed in Sections 3 and 4 and is clearly an angular unconformity. In the Carboniferous the beds dip more steeply northwards than in the Permian and this strongly suggests that higher Carboniferous strata occur offshore, beneath the Permian unconformity. All lithologies, except the evaporites and associated laminated carbonates are commonly fossiliferous, though the fauna recorded is the specialised one typical of the 'Cementstone facies' in other parts of Northern Ireland, Scotland and northern England.

Apart from changes in colour, variations in lithology can also be detected throughout the Ballycultra Formation.

1 Sandstones are thinner and less common upwards.

2 The dominant lithology in Division A and the lowest third of Division B is a micaceous, reddish and greenish grey siltstone and silty shale, but upwards these are replaced by finer dark grey shales and mudstones.

3 Though thin calcitised evaporite beds first occur immediately above the Craigavad Sandstone they are most common and reach their thickest development, in the lower half of Division B.

4 The lower half of Division C is characterised by a series of grey to yellowish grey, part dolomitic, stromatolitic micrites with algal nodules. Oncolitic limestones are absent from the other divisions.

These variations are a result of the slow environmental changes.

Depositional environments of the Ballycultra Formation

The sediments of the Ballycultra Formation bear striking similarities to tidal flat environments which have been recognised in other parts of the British Carboniferous. The presence of the following features indicate deposition near mean sea level but with minor fluctuations above and below that datum:

1 The common occurrence of desiccation cracks and ripple-marks.

2 The original existence of gypsum and/or anhydrite and possibly halite.

Plate 11 Ballycultra Formation. Interbedded grey shales and mudstones with harder ribs of micrites and calcilutites (cementstones). Foreshore west of yacht club slipway, Cultra   [NI 388]

3 The occurrence of laminated micritic dolomites or dolomitic limestones, generally without fossils.

4 The presence of algal stromatolites, serpulid reefs and stick beds.

5 The alteration of thin evaporitic beds with fossiliferous shales.

6 The thin-bedded, cyclic nature of the strata.

The desiccation cracks, algal features and the evaporites could be due to deposition in very shallow water under arid or semi-arid climatic conditions. Such features are widespread in the 'Cementstone facies' in the British Isles and evaporites are known from similar lithologies in the East Doura Borehole, Ayrshire (Bailey, 1930, p. 62) and Ballagan, Stirlingshire (Bailey, 1924, p. 11); from the Clogher Valley, County Tyrone (Sheridan and others, 1967, Sheridan, 1972, Griffith and others, in press); from Draperstown, County Londonderry (Cameron and Old, in press), Armagh area (Mitchell, in press); from the Eyam Borehole, Derbyshire (Dunham, 1973) and in the Bristol–South Wales area (W. H. C. Ramsbottom, personal communication).

The remarkable similarities of the Cultra sediments and fauna to those in counties Tyrone and Londonderry and in Scotland suggest that in early Carboniferous times similar depositional conditions preceded the first truly marine incursion throughout these areas. All these areas lie within the Midland Valley of Scotland and its westward continuation into Ireland (Belt, Freshney and Read, 1967).

The widespread distribution of strata of the 'cementstone facies' was probably reduced by later Armorican folding and erosion, prior to the deposition of Permo-Trias. This was suggested by Wright (1925) and is supported by the deep borehole at Langford Lodge, where the Permo-Trias rests directly on Lower Palaeozoic rocks (Manning and others, 1970, pp. 199–200).

The Cultra beds have many features in common with the tidal flat evaporitic facies described from the upper Viséan in Leitrim, Cavan and Fermanagh (West and others, 1968). Notable differences are the absence at Cultra of undoubted marine shales, which are a characteristic feature of part of the upper Viséan evaporitic cycles, and also at Cultra a fairly abundant specialised fauna occurs in the sandstones and greenish grey shales, mudstones and siltstones, lithologies which are almost entirely devoid of shells in the Leitrim area. These discrepancies are undoubtedly due to the differences in extent of the tidal flats, nature of the surrounding country, and also to the different positions of the two areas with respect to the tidal flat margin. The Cultra area located near the southern edge of the flat, was subject to substantial influxes of fresh water from the Lower Palaeozoic massif which tended to neutralise the hypersalinity, caused by the evaporation of shallow water, and made the environment less hostile for colonisation by *Modiolus*, ostracods, annelids, etc. True upper marine environments, if they existed, must have lain well outside the Cultra area. The Viséan tidal flats, on the other hand, are thought to have covered hundreds of square kilometres at a time, when the Connaught Coalfield lay well away from the margin and the low relief of the surrounding country, ensured that little fresh water reached the area. The effect of minor sea-level changes on such a flat area would be very great; a slight rise flooding the entire area with sea water rich in marine organisms.

Although cyclicity is seldom perfectly developed in the Cultra beds it is evident from the repetitive nature of the various lithologies. Some of the beds, notably the compact, non-laminated micrites and evaporites, commonly form double features which are, undoubtedly, due to the effect of symmetrical cycles as opposed to rhythms. Cycles are also the rule in the Connaught Viséan evaporites, although there they are developed more fully.   AB, EJC

PALAEONTOLOGY

The fossils that have been identified from the Carboniferous of the Cultra Shore are listed in Appendix 1 with the details of the sections. A full list of fossils with the authors of the specific names and the beds from which each has been collected is given in Appendix 2. The algae have been named by Dr F. W. Anderson, the ostracods by Dr J. E. Robinson and the remainder of the fossils by M. Mitchell with some bivalves confirmed by R. B. Wilson.

With the exception of the sparse faunas at the top of the succession, which are clearly linked to those in the overlying beds, the Craigavad Sandstone is unfossiliferous. The Ballycultra Formation, however, contains restricted fossil assemblages of bivalves, ostracods and annelids with associated algal layers but, except for the ostracods, these give no indication of age. The Cultra Carboniferous is chiefly noted for the well preserved ostracod faunas and was one of the main localities which provided material for the important paper on Irish Lower Carboniferous ostracods by Jones and Kirby (1896).

Dr Anderson has identified *Girvanella ducii*, *Ortonella furcata*, *O. kershopensis* and *O. tenuissima* from the algal nodules in Section 4, Bed d, and has suggested that the environment for these specimens was most likely intertidal. Several sops show desiccation cracks and many of them are bored in the outer layer (see Anderson *in* Day, 1970, p. 175).

The commonest fossils are plant fragments, *Modiolus latus*, ostracods, the worm tubes *Serpula sp.* and *Spirorbis sp.* and fish fragments. Some of the specimens of *Modiolus latus* have a *Curvirimula*-like appearance which is due to their crushed state of preservation in smooth grey mudstone. *Spirorbis sp.* is usually preserved attached to bivalve fragments, but *Serpula sp.* often occurs as compact interwoven masses of tubes (up to $55 \times 65 \times 75$mm). In this form it is possible that these worm tubes were mistaken for the narrow corallites of the rugose coral *Lithostrotion junceum* which was recorded by Wright (1872, p. 36). *L. junceum* has not been found in the present collections from Cultra.

The *Modiolus*-ostracod faunas, with an abundance of individuals of a small number of species, suggests a specialised brackish or, more possibly in view of the presence of evaporite horizons, hypersaline water environment. More marine conditions are probably indicated by the rare records of rhynchonelloids including 'Camarotoechia' sp. and indeterminate gastropods. 'Estheria' has also been recorded (see below).   MM

The only occurrence of 'Camarotoechia' in Division B is in a nodule at the base of Bed 63, an olive-grey micrite also containing 'Estheria'. The brachiopods are associated with abundant large fragments of fish in such a way as to suggest that the fish skeleton served as an anchoring point in an otherwise muddy environment. If this is so then the distribution of 'Camarotoechia' may be controlled more by the nature of bottom sediments than by salinity.

Of particular interest is the occurrence of 'Estheria' in Bed 63 where some bedding planes are crowded with this poorly known thin-shelled, brachiopod crustacean. 'Estheria' makes rare appearances in the intertidal evaporitic deposits and is known from:

(a) The Cementstone Group of Killock Burn, North Ayrshire (Bailey, 1930, p. 61).
(b) The Meenymore Formation of Slieve Rushen and Curteagh (Brandon, 1977). Other occurrences of 'Estheria' where hypersalinity in associated sediments has not been proved but could be suspected are in the Upper Oil Shale Group of the Edinburgh region at Rosyth (Mitchell and Mykura, 1962, p. 74; Anderson, 1950, p. 22) and at Livingston and Pumpherston (Carruthers, 1927, pp. 44, 56).

The limestones and shales with abundant *Serpula* tubes in Division C can be compared with the serpula biostromes and reefs described recently from the Lower Border Group of Cumberland (Leeder, 1973) for which there is an analogue in the serpulid reefs of Connemara, Eire (Bosence, 1973). MM, AB, EJC

Ostracods

Two lines of information emerge from the study of shale samples from the Cultra Carboniferous. First, the fauna of ostracods allows certain palaeoecological conclusions to be drawn. Second, a correlation can be made with sequences in northern England on the occurrence of several species. The more local correlation of the fault-separated blocks of the foreshore is less easy. From a study of the local vertical distribution of ostracods within the sections it can be ascertained that the basal 4.57 m of Section 1 overlaps with the upper part of Section 2 and that a similar top-bottom overlap may relate the base of Section 4 to the top of Section 5. Both latter sections may be younger than Section 1. These conclusions are in close agreement with those resulting from correlation using evaporite marker bands and detailed stratigraphical studies.

Facies and palaeoecology

In the ecology of recent muds and clays of estuaries and marshes, it is frequently observed that when salinities fall into the brackish-water range, especially values of between 5 to 10 parts per thousand, ostracod assemblages are made up of only a single, or at most two species, of which there may be many individuals. In more saline waters, the diversity of species increases markedly to what might be called a 'full marine fauna'. Similar diversity marks the occurrence of true fresh-water environments. Restriction on the number of species would also characterise a hypersaline environment. Short of palaeosalinity determinations upon containing sediments, this diversity pattern is as close as one can get to using fossil ostracods as palaeoenvironment indicators, coupled with whatever information can be obtained from the associated fauna.

Upon this basic assumption, many of the shales from Cultra could be judged to have been deposited in high or possibly low salinity waters on Carboniferous tidal-flats, an environment also suggested by the algal-growths, spirorbid and serpulid worm tube concentrations, and certain sedimentary structures. In particular, those samples with ostracod assemblages dominated by *Acutiangulata* cf. *sublunata* or *Cavellina* cf. *extuberata*, could represent restricted waters (see Section 1, Beds f to h; Section 5, Beds f and j). By the same token, assemblages with five or six genera, and a total of six to eight species would represent more marine waters (see Section 1, Bed a; Section 1, Bed h; Section 5, Beds e and h). The marine shales of Cultra, however, contain fewer elements than the marine shales of the Roxburghshire Cementstones Group, with which they otherwise compare in age. Just as there are no chonetoids, productoids or syringothyrids at Cultra, so there are no species of *Bairdia*, *Bythocyproidea*, *Hollinella*, *Kirkbya* or *Quasillites* among the ostracods.

The overall character of the Cultra sections suggests a depositional area less directly connected to the open sea than the Liddesdale 'Gulf' to the south of the Southern Uplands. The conditions, however, could be matched in the Cementstone Group beds of the Kielder area of the North Tyne, or in the Coquet Valley to the north.

Correlation with other Carboniferous areas

The facies assemblages mentioned as indicating restricted conditions, range through much of the Lower Carboniferous, but it is possible to distinguish an Upper Border Group (=Scremerston Coal Group of Northumberland) age from a Lower Border Group age for assemblages from the same facies. Both would consist of the same genus, *Acutiangulata*, but the species would be different. The Cultra assemblages compare best with those from the Lower Border Group of Liddesdale. Similar conclusions follow from consideration of the more marine species. *Acutiangulata quadrata*, *Beyrichiopsis annectens*, *B. fortis glabra*, *B. plicta* (s.s.), *Cavellina incurvescens*, *Cavellina spola*, *Cavellina* cf. *taidonensis*, *Glyptolichwinella annularis* and *Knoxiella* cf. *rugulosa* are represented in the beds above and below the Main Algal Reef of Garwood, 1931 (=MA12 of the Main Algal Beds, Lower Border Group, of Day 1970, pp. 43–66) in Roxburghshire and northern Cumberland.

Acutiangulata sublunata, *Beyrichiopsis plicata* and *Cavellina extuberata* were all recorded from the Carboniferous strata penetrated by the Gayton Borehole in Northamptonshire. The ostracod faunas occurred (Jones and Kirby 1886a, p. 249; Eunson, 1884, p. 487) from 730 to 731 ft (222.5 to 222.8 m) in this borehole, in limestones and shales which have been equated with the Lower Limestone Shale of the South-west Province.

Most of the ostracod species mentioned do not range up into the Scremerston Coal Group of Northumberland and no typical Scremerston ostracods were found at Cultra. The similarities with the Lower Border Group faunas of Liddesdale on the other hand are impressive.

Age of Lower Carboniferous strata at Cultra

Dr Robinson's work on the ostracod faunas (see Appendix 2) clearly indicates a correlation of the Ballycultra Formation with part of the Lower Border Group of Liddesdale (Day, 1970, p. 10). The lack of clearly diagnostic macrofossils makes the dating of these lowest beds in the Northumberland Trough difficult but they have been generally regarded as being of late Tournaisian or early Viséan age. George and others (1976) referred these lowest Carboniferous beds (p. 41) and the beds exposed at Cultra (p. 58) to the Courceyan Stage. However, recent work on the conodonts and foraminifera from the Lynbank Beds (basal Lower Border Group of Bewcastle) quoted by Ramsbottom (1977, p. 283 footnote) indicates that the basal beds in the Northumberland Trough are of Chadian age and if the correlations are correct then the Cultra succession is also of this age. MM

PETROGRAPHY AND DIAGENESIS (EVAPORITES)

Former anhydrite and gypsum beds have been extensively, or completely, calcitised through the circulation of groundwater with bicarbonate ions and are now represented by thin, pink or cream-coloured sparry secondary limestone beds up to 8 cm thick. The calcitised evaporites are almost invariably developed above, or in the top of, thin beds of fine-grained, laminated, olive-grey, micrite from 1 to 3 cm thick. The micrites are normally unfossiliferous except for carbonaceous plant remains, although broken ostracod tests are present occasionally. Lenses of quartz-sand are common and fine banding in some of the micrites may indicate a stromatolitic origin. Many of the relict evaporites are nodular, ie the secondary limestone can be discontinuous and bedded in a nodular form, but all evaporites persist laterally and are thus useful 'marker bands'.

As noted earlier the evaporites are not uniform in distribution throughout the Ballycultra Formation. Only three are known from Division A. These are: Evaporite C, about 31 m above the base of the Ballycultra Formation and near the top of Section 1, Appendix 1. This nodular evaporite could not be placed accurately in this section. Evaporite B, about 15 m above the base of the Ballycultra Formation. Evaporite A, 1.8 m above the base of the Ballycultra Formation.

The evaporites are most abundant in Division B and 14 occur in this division. Stratigraphical details are given in Appendix 1. One thin calcitised evaporite occurs 14.24 m above the base of Division C.

The mode of occurrence and texture of the Cultra evaporites bear striking similarities to those described from the Purbeck of England (West, 1964, 1965) and from the Viséan of Ireland (West and others, 1968). Other examples of calcitised evaporites are given by West (1973). That the secondary limestones of the Ballycultra Formation are calcite replacements of anhydrite and sometimes gypsum is proved by the presence of the following relict evaporite textures and diagenetic replacements:

		Evaporites studied with these features %
(a)	Macrocell texture[1] (frequency called nodular texture)	71
(b)	Net texture	50
(c)	Enterolithic veining or folding (West, 1965, p. 49)	59
(d)	Lath-shaped anhydrite ghosts in an anhedral calcite mosaic	75
(e)	Calcite pseudomorphs after prismatic anhydrite	83
(f)	Calcite pseudomorphs after gypsum	17
(g)	Secondary lutecite (length-slow chalcedony)	8
(h)	Secondary quartz	8

As yet celestine (strontium sulphate), which commonly replaces gypsum and anhydrite (West, 1973) has not been detected in the Cultra evaporites. Twelve of the evaporites have been studied in this section. The details of these, and the remaining evaporites, are given in Table 2.

The stages of diagenesis of the Purbeck calcium sulphates involving a change from gypsum to anhydrite and then back to gypsum (West, 1964, 1965) would appear to have been followed by the Cultra evaporites.

The Purbeck stages are as follows:

Stage I   Syngenetic, primary, discrete, lanceolate crystals of gypsum.
Stage II   Anhedral gypsum with net-texture (IIa) and macrocell texture (IIb). Net-texture arose from the compaction, and perhaps crystal growth, of Stage I gypsum lenses, followed by the conversion of the mass into anhedral gypsum with considerably reduced porosity. Enlargement of individual units in the net-texture produces macrocell texture and enterolithic veins.
Stage III   Net-texture (IIIa) and macrocellular texture (IIIb) filled with a matrix of lath-shaped anhydrite. This stage arose from Stage II by a loss of water of crystallisation. The diagenesis of the macrocell texture forms a parallel sequence to that of net-texture.
Stage IV   The growth of gypsum porphyroblasts within Stage III anhydrite by the hydration of anhydrite with a limited supply of water.
Stage V   This differs from Stage IV in that the matrix between the gypsum porphyroblasts has also been hydrated. Satin-spar (Stage Vc) is developed by displacement in the anhydrite during hydration.

In the case of the Cultra evaporites, however, calcitisation was in progress during Stages I to III and complete calcitisation must have occurred by the end of Stage III, effectively arresting sulphate diagenesis. Calcitised replacements of Stages IV and V (and in particular satin-spar veins) do not occur in the evaporites at Cultra.

Stage III is represented by the majority of beds at Cultra. Stage II has been preserved by the calcitisation of net-textures, with gypsum lenses, on only two levels. West's Stage I of early gypsum lenses is represented by small isolated calcite pseudomorphs after gypsum in micrite, associated with calcitised Stage II gypsum.

[1]The term macrocellular is used when it is applicable to an internal evaporite texture, ie a secondary limestone nodule can be composed of any number of macrocells.

Table 2 Occurrence of relict evaporite textures and of diagenetic replacements of evaporites at Cultra, Co. Down

Evaporite division	A			B															C
	A	B*	C	A(i)	A(ii)	B	C	D	E	F	G	H	I	J	K	L	M	N	C
Thin section (NI)	—	3728	—	3729	—	3730	3731	—	3732	3733	3734	3735	3736	3737	3738	—	3739	—	—
Thickness (mm)	40	50	40	30	4	20	50	50	80	14	12	20	40	30	70	20	30	25	50
Colour	P	P	P	P	P	P	P	P	P	P	C	P	P	C	C	P	P	P	P
Nodular bedding	.	.	X	.	.	X	X	X	.	X	.	.	X	X	.	X	X	X	.
Macrocells	X	.	X	.	.	X	.	X	X	X	.	X	X	.	X	X	X	.	X
Enterolithic veining	X	.	X	.	X	.	X	X	X	.	X	.	X	.	X	–	.	.	X
Net texture	–	X	–	X	.	.	.	–	.	.	X	X	.	X	X	–	.	X	–
Ghosts of anhydrite	–	X	–	X	.	X	.	–	X	X	.	X	X	X	.	–	X	–	–
Pseudomorphs after anhydrite	–	X	–	X	.	X	X	.	X	X	.	X	X	X	.	–	X	–	–
Pseudomorphs after gypsum	–	.	–	.	.	.	.	.	.	.	X	.	.	.	X	–	.	–	–
Lutecite replacement	–	X	–	.	.	.	.	.	.	.	.	.	.	.	.	.	.	–	–
Quartz replacement	–	.	–	.	.	.	.	.	.	.	X	.	.	.	.	.	.	–	–
Dolomite	–	.	–	.	.	.	X	–	.	.	.	.	.	.	.	–	†	–	–
Diagenetic stage	–	IIId	–	IIId	.	IIIb	III	–	IVb	IIIb	I+IIa	III a/b	IIIb	IIIa	I+IIa/b	–	IIIb	–	–

A feature observed is indicated by X, and not observed by a dot . ; where no thin section was
available and the presence of some characters could not be clearly determined, dashes – are
inserted. Letter P indicates colour pink, C cream-coloured. Most such rocks are probably of Stage III b.

* Evaporite B, of Division K was taken adjacent to a basalt dyke on the foreshore at Section 1,
90 m NW of Duaraltic House. The strata in the vicinity of this dyke have been indurated.

† Occasional.

Dolomite is rare in replaced evaporites (West, 1973) but this mineral occurs as abundant euhedral rhombs throughout Evaporite C, Division B, and is occasionally found in Evaporite M, Division B.

Stage I and II evaporites

These stages are preserved on only two levels, ie Evaporites G and K, Division B. Evaporite G consists of a 12 mm cream-coloured layer of net-texture at the top of a 30 mm bed of laminated micrite. The net cells are filled with a mosaic of anhedral calcite with anhedral grains of secondary quartz (mega-quartz) up to 0.05 mm across. The micritic cell walls are thin and often indistinct. In the micrite below the evaporitic layer there are isolated cells and small lanceolate calcite pseudomorphs after gypsum. Calcite pseudomorphs after anhydrite are absent from this rock. The lowest part of the evaporite bed is sometimes strongly folded.

Evaporite K is a prominent cream-coloured bed, 7 cm thick, resting on a thin laminated micrite bed up to 1 cm thick. The evaporite layer is strongly enterolithically folded. This rock contains both macrocells and net-texture and the micritic material caught up in the folds and composing the thin cell walls contains micritised lens-shaped calcite pseudomorphs after gypsum. The rock is therefore midway between Stage IIa and b.

Stage III evaporites

Calcite in these cases replaces anhydrite. Pseudomorphs after, and relict 'ghosts' of, anhydrite prisms are characteristic of these rocks and there is a complete absence of corresponding gypsum relics. Both Stage IIIa, with net-texture, and IIIb, with macrocellular or nodular texture, are developed. Some levels contain both textures and represent intermediate stages. This feature is also found in the Purbeck specimens (West, 1965). With only one exception Stage III calcitised evaporites have a distinct pink colour.

(a) Stage IIIa Calcitised net anhydrite. This is represented by Evaporites A and J of Division B and Evaporite B of Division A. Enterolithic folding is absent or very poorly developed in these specimens. The most interesting example is Evaporite B, Division A (Plate 12). Two 2-cm beds of net-texture are developed in a thin laminated micrite. The cells are filled, or partly filled, with an anhedral calcite mosaic with 'ghosts' of lath-shaped anhydrite. Many of the cells contain spherules of lutecite. This mineral is only found in evaporite replacement beds and is a useful criterion for detecting vanished evaporites (Folk and Pittman, 1971; West, 1973). The thin micritic cell wall material contains good calcite pseudomorphs after prismatic anhydrite.

(b) Stage IIIb   Calcitised macrocellular anhydrite. These evaporites ranged from those consisting of solitary macrocells in a micrite bed (eg Evaporite F, Division B) to relatively thick, massive beds composed of large macrocells with thin cell walls. A typical example of the latter, and the best exposed, is Evaporite E, Division B (Plate 12). This is a fairly continuous, pink bed, 8 cm thick, becoming nodular only on occasions. It rests on 2 cm of laminated, olive-grey micrite with quartz-sand lenses. The large macrocells (up to 4 cm across) have a tendency to form enterolithic veins with micrite caught up between the folds. This folding or veining takes the form of a series of domes and basins giving rise to a hummocky or mammillate surface (cf. West, 1965), pl. 3, fig. 1). The base of the evaporite layer is also irregular. The anhedral calcite mosaic of the macrocells displays an evaporite fabric with abundant relict ghosts of anhydrite prisms. Calcite pseudomorphs after anhydrite protrude from the macrocells into the thin micrite walls and into the micrite layer below. Many small irregular hollows in the calcite mosaic may be due to anhydrite, or a more soluble replacement than calcite, being removed in solution.

Period of calcitisation

The diagenetic history of calcium sulphate in the Purbeck and Cultra deposits involves changes in water content (either of pore water or water of crystallisation) which are likely to have been dependent on variations in the weight of overburden (West, 1964, p. 323). During Stage I to Stage

A

B

C

D

Plate 12   Evaporites from the Ballycultra Formation

A. Calcitised net-textured gypsum Stage IIa Evaporite G, Division B

   Cream-coloured calcitised gypsum. The layers of net-texture with some secondary quartz have been enterolithically veined and occur above a thin laminated micrite bed with isolated cells. × ⅔

B. Calcitised net and macrocellular gypsum   Stage IIa/b   Evaporite K, Division B

   Cream-coloured enterolithic veins and macrocells of calcitised gypsum resting on thin laminated micrite and passing up into net texture. The thin cell walls contain micritised pseudomorphs after gypsum. × ⅔

C. Calcitised net-texture anhydrite   Stage IIIa   Evaporite B, Division A

   Thin pink beds of net-textured anhydrite alternate with laminated sandy micrite. Staining with alizarin red has darkened the sparry calcite net-filling but left white the lutecite chalcedony. The thin micritic cell walls contain good calcite pseudomorphs after prismatic anhydrite. × 1

D. Calcitised macrocellular anhydrite   Stage IIIb   Evaporite E, Division B

   Macrocellular calcitised anhydrite rests on thin laminated carbonate. The macrocells are filled with a pink porous calcite mosaic and are enterolithically veined and folded. Pseudomorphs after anhydrite are abundant. × ⅔

III this overburden must have been increasing while the change from Stage III to Stage IV, and eventually Stage V, required a considerable decrease in overburden allowing hydration to occur. From theoretical considerations (Macdonald, 1953) and evidence of other sulphate deposits (Ogniben, 1957) it seems that the change to anhydrite (ie Stage II to Stage III) took place under an overburden of not more than 600 m. It is evident that calcitisation (produced by circulating groundwater charged with bicarbonate ions) was completed by the end of Stage III, otherwise some of the remaining anhydrite would have been hydrated during subsequent reduction in overburden. With Cultra situated on the southern edge of the continuation of the Midland Valley of Scotland it is most unlikely that the Ballycultra Formation was not overlain by more than 600 m of later Carboniferous sediments. This means that complete calcitisation must have occurred during Carboniferous times otherwise hydration to Stages IV and V would have occurred during erosion before the deposition of the basal Permian rocks. (The completely calcitised evaporite in Division C is only 6.7 m below the Carboniferous–Permian unconformity.)

The two Stage II evaporites must have been calcitised before sufficient weight of Carboniferous strata had accumulated. This preferential replacement process is not clear and it is not known if calcitisation was a continual process or occurred on two separate occasions within the Carboniferous period. The reason for the difference in colour between Stage II calcitised evaporites, which are cream, and Stage III calcitised evaporites, which with one exception are pink, is also not clear. The calcitised macro-cell beds from the Altagoan Formation of the Carboniferous, Altagoan River, Draperstown, Co. Derry (Sheet 27) are also partly pink.   AB

CONCEALED COALFIELD NORTH OF CULTRA

Ever since Wright (1925) postulated a concealed coalfield in the North Channel off the south-east Antrim coast, around Larne and Island Magee, there has been no evidence forthcoming of its existence. Recently, however, Carboniferous miospores have been extracted from a sample of variegated red and green Triassic mudstone from a borehole at Palace Army Barracks. The Carboniferous miospores, identified by B. Owens, indicate a Namurian–Westphalian age and include:

Apiculatisporis sp., Dictyotriletes sp., Tripartites trilinguis (Horst) Smith & Butterworth (1967), *Lycospora pusilla* (Ibrahim) Somers, *Densosporites anulatus* (Loose) Smith & Butterworth, 1967, *D. pseudoanulatus* Butterworth & Williams (1958), *Laevigatosporites sp.*

Accompanying these miospores are sporadic Permo-Triassic bisaccate miospores which indicate that the Carboniferous specimens are re-worked. The abundance of derived Upper Carboniferous miospores suggests that rocks of this age lie beneath a Permo-Triassic cover not too far away, probably in the region indicated by Wright and partly within the confines of the present sheet under Belfast Lough.   AB

CHAPTER 5

Permian

INTRODUCTION

Rocks of known Permian age crop out in Northern Ireland in two areas; at shore sections at Cultra [412 809] and in Tyrone (Fowler and Robbie, 1961). A borehole at Avoniel in east Belfast has proved a concealed Permian succession beneath the city and boreholes have shown that this extends up the Lagan Valley to beyond Long Kesh (Manning, Robbie and Wilson, 1970). Permian rocks are also known to occur in the Newtownards area.

In late Westphalian times intermittent earth movements took place and continued possibly into the Permian. Thus, in most of the north of Ireland, the Lower Permian was mainly a period of subaerial erosion during which, the newly raised areas were deeply dissected and debris from the hills accumulated in fans at mountain fronts and in subsiding intermontane troughs. These sediments in Northern Ireland were coarse water-lain breccias. The thick basal sandstones and brockrams of the Belfast (129 m) and Newtownards areas (300 m) are represented at Cultra by only 1.5 m of a polymict brockram which may be a residual piedmont gravel.

The Upper Permian was ushered in by the widespread transgression of the Bakevellia Sea (Smith, D. B., 1970), arms of which penetrated parts of eastern Ireland. In Northern Ireland the first marine deposit was the dolomitic Magnesian Limestone which has yielded a sparse bivalve-gastropod fauna of early Upper Permian age. It is about 8.5 m thick at Cultra and other proven thicknesses are 21 m at Avoniel and 18 to 23 m in Tyrone. The fauna of the carbonate rocks is a key factor in the correlation of various marine sequences and has been reviewed by Pattison (1970 and *in* Manning and others, 1970, pp. 30–31).

The Magnesian Limestone is succeeded by red mudstones (the Permian Upper Marls) which outcrop at Cultra and do not contain gypsum or anhydrite as at Avoniel and other localities. These mudstones are succeeded by sandstones of 'Bunter' facies, now assigned to the Sherwood Sandstone Group. Recent spore evidence from Kingscourt, Co. Cavan (about 100 km SSW of Belfast) suggests that the Permian–Triassic boundary may lie within this mudstone sequence, but, as in the Belfast area (Manning and others, 1970, p. 31), those deposits are all considered in this chapter, although on the published map (Sheet 29) the Cultra occurrence is erroneously assigned to the Triassic.

The exposed succession at Cultra is:

	Thickness m
Permian Upper Marls	3 +
Magnesian Limestone	8.43
Breccia with Carboniferous and Lower Palaeozoic fragments	1.5
Unconformity	
Lower Carboniferous	

There is an extensive literature on the Permian at Cultra, but because the Permian is so intimately associated with the Carboniferous the two systems have often been confused and the literature is correspondingly muddled.

The strata were first designated as Permian by Bryce (1837b) but it is evident, both from his map and the text, that the rocks which he was considering as Permian were the thin cementstone ribs interstratified in the Carboniferous shales. These cementstones are dolomitic and it is because of this and their close association with red-beds that Bryce regarded them as being of Permian age. Bryce also noted minute terebratulids or productids but it is clear that these were obtained from what we now know to be Carboniferous strata. At first Griffith (1837, p. 148) supported Bryce's correlation and it is evident that he, like Bryce, considered that all the strata exposed on the Cultra foreshore belonged to one system. Later, Griffith changed his opinion and stated (Griffith, 1843, p. 45) that he considered the strata present on the foreshore at Cultra to be of Carboniferous age.

However, Bryce maintained their Permian status and in 1852, on the occasion of the British Association meeting in Belfast, published two papers. In one of them (1852, pp. 21–22) he evidently considered the whole foreshore at Cultra, including Carboniferous and Permian strata, to be of Permian age but in the other (1853, p. 42) he inferred that both Carboniferous and Permian were present at Cultra. At this time Binney was staying in Belfast and he visited the Cultra foreshore where he was able to distinguish both Carboniferous and Permian strata (Binney, 1855). It is clear from the text that Binney recognised the true Permian containing *Schizodus* and *Bakevellia* at its only known outcrop and therefore the credit for identifying the Permian Magnesian Limestone at Cultra, without any ambiguity, belongs to him. He made no map, however, and the relationships of this formation to the neighbouring strata remained undefined.

Bryce (1852, p. 22) repeated the record of terebratulids and productids from the Permian of Cultra previously mentioned in his 1837 paper. Since at the time of writing Bryce had not differentiated Permian from Carboniferous strata it is possible that they were collected from Carboniferous rocks. However, brachiopods are exceedingly uncommon both in the Carboniferous and the Permian strata. This fact is brought out by King (1857, p. 79) where, in stating that he thought Bryce had made an error, he noted that the fossils did not occur in Griffith's 1843 list which had been compiled from both the Permian and Carboniferous. This mention of the occurrence of brachiopods in the Permian of Cultra is worth noting, in spite of its ambiguity, on account of the later record of *Productus horridus* (Baily *in* Hull and others, 1871). This discovery has never been repeated since, in spite of detailed collecting during the present resurvey, and it is now evident that the presence of brachiopods in the Magnesian Limestone at any Irish locality is a distinct rarity.

Also, in 1852, King (p. 53) published a list of Permian forms found at Cultra consisting of *Schizodus schlotheimi*, *Pleurophorus costatus*, *Bakevellia antiqua*; and MacAdam (1852, p. 53) also recognised *Schizodus*. Subsequently, King figured a number of fossils mainly from Tullyconnell in County Tyrone but noted (1857, pp. 76–77) that the records of *Cucullea unilateralis*, *C. complanata*, *C. trapezium* and *C. hardingii* given by Griffith from the Cultra locality (1843, p. 46) were probably misidentifications of *Schizodus schlotheimi*.

The Permian and Carboniferous strata at Cultra were first systematically mapped by G. V. Du Noyer in 1867, but the description was not published until 1871 because of his death. In the Memoir the Permian was accurately separated from the Carboniferous and a detailed description of its exposure was given (Hull and others, 1871, pp. 10–11 and 20.). In particular, the fact that the Permian is unconformable to the Carboniferous was established. The palaeontological work was done by W. H. Baily (*in* Hull and others, 1871, pp. 18–19) who recorded *Schizodus schlotheimi*, *Bakevellia antiqua*, *Arca*-like *tumida*, *Productus horridus* and *Turbo helicinus*. The brachiopods and the gastropods were remarked to be 'not so abundant'.

After the publication of the Memoir there was a further period of confusion between the Carboniferous and Permian strata of Cultra which has been discussed in Chapter 4.

The area was resurveyed during the preparation of the Belfast Memoir (Lamplugh and others, 1904, p. 18). Because of the confusion which had arisen over this section, Lamplugh revisited the area at low water of Spring tide and measured the section. His measurements do not entirely agree with those recorded during the present resurvey.

Lamplugh's section

Permian   Ft   (m)

Flaggy, dull red and slightly mottled sandy, slightly gritty, marls, with occasional ripple-marked layers. Thickness seen   about 10 (3.05)

Red, gritty, and fine pebbly marl—a fine-grained 'brockram'—the fragments chiefly crumbs of shale and sandstones, but with a few subangular bits of quartz towards the base (the largest observed being 1.5 inches in diameter) where the bed is coarser, and yellowish or greenish; the whole thins out westwards and is probably a lenticle   2 (0.61)

Magnesian Limestone

Band of yellow magnesian limestone, mostly soft and decomposed, but in places hard and showing traces of fossils; lumpy at the top as if nodular and waterworn   1.5–2.5 (0.48–0.76)

Flaggy dolomitic yellow and reddish purple marly layers, impersistent and probably lenticular. Greatest thickness not more than   2 (0.61)

Gritty, fine-textured 'brockram'; the fragments chiefly of sandy shale and sandstone, but with some angular and subangular pieces of quartzite, vein-quartz, slate and chert (the largest observed, 3 inches in largest diameter); irregular in thickness

and probably lenticular; resting unconformably on stained Carboniferous shales   about 2.5 (0.76)

The two uppermost units agree well both in thickness and description with the section measured during the resurvey (see Appendix 1) as does the brockram. The discrepancy is in the thickness of the Magnesian Limestone; 4.5 ft (1.37 m) recorded by Lamplugh and 8.43 m in the present resurvey. This cannot be explained by suggesting that Lamplugh's section was measured in a slightly different place and that on account of the highly faulted character of the outcrop different portions of the succession were measured and arranged in a composite section. It seems possible that there was some error in transcription, particularly in view of the accuracy with which Lamplugh recorded the other beds.

Particular care was taken to check Lamplugh's section and the foreshore was visited at low water of Spring tide to ensure that the most complete section was exposed. It was established that all the visible Permian had been seen and measured and Lamplugh's observation was also confirmed that the beds of marl (Permian), which overlie the Magnesian Limestone, are themselves overlain by boulder clay which was seen to extend seawards for about 50 m before being covered with mud.

The systematic position of the Permian strata at Cultra has always been a matter of debate. King (1857, pp. 78–79) considered that the Tullyconnell (County Tyrone) and Cultra occurrences were laid down in the same basin of deposition. He noted the lithological and chemical similarities of the two areas and, in observing that the fossil assemblages were identical, compared them to the Upper Magnesian Limestone of Durham.

In making this correlation, King was possibly influenced by the absence of brachiopods from the Irish fauna for he specifically stated (1857, p. 79) that Bryce's record of brachiopods (1852, p. 22) was probably a mistake. King also compared the Irish Permian to the Magnesian Limestone at Whitehaven in Cumberland (1857, p. 80) which he regarded as linking the Irish Permian to the Upper Magnesian Limestone of Durham.

The 1871 Memoir repeated King's diagnosis of these beds as the equivalent of the Durham Upper Magnesian Limestone (p. 11) although the characteristic Lower Magnesian Limestone form, *Productus horridus*, was recorded. In the 1904 Memoir the systematic position of the Cultra beds was not discussed but the similarity in the successions at Cultra and at Whitehaven was stressed (Lamplugh and others, 1904, p. 18).

In his correlation of British Permo-Triassic rocks Sherlock (1926, pp. 44–46) considered that 'the Magnesian Limestone of Ireland is of the same age as that of Whitehaven, and most probably that of south Lancashire'. He also noted the record of *Productus horridus* from Cultra and stated (1926, p. 33) 'the presence of *Productus* at Cultra in strata generally considered to be the same age as the Whitehaven limestone is strong evidence against the strata being the Upper Limestone'. Thus, Sherlock used the record of *Productus* from Cultra not only to establish the Lower Magnesian Limestone age of the Irish Permian but also of all the Permian strata west of the Pennines as well.

Sherlock's argument was restated by Stubblefield, *in* discussion of Dunham and Rose (1949, p. 40) where he wrote, 'therefore, whatever the age of the Lancashire and Cumberland Magnesian Limestone, if the *P. horridus* record was reliable, the County Down fauna should continue to be regarded as of Lower Zechstein age'.

Consequently, the suggestion of Fowler (1955, p. 52) that the Tyrone Permian and, by implication, the Cultra Permian is equivalent to the Upper Zechstein of Durham, without reference to the intervening Permian strata of north-west England, is surprising. This interpretation depends on the presence in the upper part of the Tyrone Magnesian Limestone of the problematical fossil *Calcinema* [*Filograna?*] *permiana*. However, this fossil is also present in the Permian of north-west England where again the systematic position of the Permian strata is not clear (Stubblefield, *in* discussion on Dunham and Rose, 1949, p. 40). Fowler's interpretation was repeated by Fowler and Robbie (1961, p. 93) in which, however, Stubblefield suggested a Lower Magnesian Limestone age for strata below those containing *Calcinema* [*Filograna?*] *permiana*. Hirst and Dunham (1963, p. 935) equated the Hilton Plant Bed of Cumberland with the Marl Slate of Durham, a correlation implying the presence of members of the Lower Magnesian Limestone in north-west England and Ireland.

More recently, Pattison (1970, p. 161) concluded that the patterns of faunal distribution suggest an approximate correlation of most of the Permian marine strata of Northern Ireland with the carbonate phase of Zechstein 1, including the Lower Magnesian Limestone of Durham. This conclusion was broadly followed by Smith and others (1974).

It is the almost complete absence of brachiopods from the Irish Permian which has, from the first, led to the successions being assigned to the Upper Zechstein, for in north-east England, where the succession is fully represented, the Lower Magnesian Limestone commonly contains brachiopods. The Zechstein environment was, however, that of an evaporite sea in which, typically, a restricted marine environment followed by evaporite deposition developed, first at the margins and later in the centre of the basins. Under these conditions a dolomite with a bivalve-gastropod fauna and possibly very occasional brachiopods at the edge of the basin could be the time equivalent of a dolomite with brachiopods in the centre of the basin. There can be no doubt that the Cultra deposit was at the extreme edge of the basin, bounded immediately southwards by the Southern Uplands ridge, and if the conditions postulated did in fact obtain then the difficulties met in correlating the Permian strata of Ireland are more readily understood.

The Irish Permian has always been correlated with the Permian of north-west England on the grounds that these are the nearest similar deposits, and show many lithological and sequential similarities. As Shotton, *in* discussion on Dunham and Rose (1949, p. 39), has remarked, 'the only satisfactory connexion between the Irish deposits and those of northern England was via a strait which crossed the southern end of the Furness district'. A strait of this kind extending north-westwards from the Furness area has been discussed in the chapter on the Carboniferous since it is also needed to account for similarities in the Carboniferous stratigraphy. Briefly, a channel through the Southern Uplands is thought to have developed along a fault line in the Lower Palaeozoic basement. The channel provided a link between different basins of sedimentation during Carboniferous times. Variscan reactivation of this and other faults broadly resurrected the pre-Carboniferous topography within which the Permian strata were deposited; the North Channel strait persisting and again providing a link between Ireland and the north-west of England through the Southern Uplands barrier. However, the connection between the Bakevellia Sea of Northern Ireland and north-west England on the one hand and the Zechstein Sea of north-east England and north central Europe on the other may well have been indirect by way of separate links to a northern ocean (Smith and others, 1974, p. 7).

BASAL PERMIAN BROCKRAM

The thickness of this formation varies even over the small distance of the visible outcrop from 0.6 to 1.5 m. It always occupies a position between the Magnesian Limestone and the underlying Lower Carboniferous strata. It consists entirely of fragments derived from pre-existing formations; sandstones and cementstones from the Lower Carboniferous and greywackes and slates from the Lower Palaeozoic outcrop. The fragments are usually angular; the average size is about 2 cm in diameter but pieces up to 20 cm have been noted. The matrix of the rock is generally purplish and the overall colour may be purplish or greenish.

The thickness seen here of 0.6 to 1.5 m contrasts sharply with the 44 m+ found in the borehole at Avoniel, 8 km to the south-west. This variation in the thickness of the basal Permian is well known elsewhere and is commonly explained as the infilling of an irregular pre-Permian surface. It is also probable that the nearby Lower Palaeozoic ridge was the chief contributor to the Permian brockram. While the deposits at Cultra are a piedmont veneer, it is probable that in areas of thicker deposition the brockrams took the form of scree slopes and outwash fans.

MAGNESIAN LIMESTONE

The Magnesian Limestone at Cultra is, for the most part, a massive, yellow dolomite with a gritty texture about 8.4 m thick. The basal 2 m is more variable and is variously spotted and banded in shades of pink, mauve, purple and black. Within the dolomite there are fossiliferous horizons but these are subordinate to the intervening masses of unfossiliferous dolomite. The fossiliferous bands are generally not more than 0.3 m thick and may vary from dolomite, carrying a few evenly dispersed shells, to beds consisting mainly of bivalves which were essentially shell banks.

The rock has a sandy, porous texture. The shells of fossils are usually dissolved away and are represented only by casts. The porous character of the rock is perhaps emphasised by the deep weathering to which it has been subjected on the wave-washed platform.

Within the exposure the dip is fairly constant at 20° to 25°

NW and the thickness appears to be constant, although this is obscured by faulting.

PERMIAN UPPER MARLS

The junction with the Permian Marls is well exposed and no angular discordance is visible; the junction itself is sharp and regular, but there are signs of irregular weathering of the Magnesian Limestone surface. There are no pebbles of dolomite in the basal beds of the mudstones.

The mudstones contain no gypsum or anhydrite, but these minerals may have been dissolved out.

PALAEONTOLOGY

King (1852) and Lamplugh and others (1904) listed several species collected from the Magnesian Limestone and further large collections from that bed are held by the Institute of Geological Sciences. The latter collections, as well as some of the earlier ones, have been examined by Pattison (1970). He recorded: *Agathammina milioloides, A. pusilla, Calcitornella? minutissima,* cf. *Cyclogyra sp.,* cf. *Hemigordius? sp.,* and gastropods including cf. *Coelostylina? leighi,* cf. *C.? obtusa, Glyptospira? helicina,* cf. *'Loxonema' phillipsi, Naticopsis minima,* cf. *Dentalium sp., Bakevellia (B.) binneyi, B. (B.)* cf. *ceratophaga, Liebea squamosa,* cf. *Nucula? sp.,* cf. *Parallelodon striatus, Permophorus costatus, Schizodus obscurus,* and ostracods and commented, that as in County Tyrone, bivalves and especially internal and external moulds of *Bakevellia (B.) binneyi* are the most common fossils. The gastropods are also fairly abundant but in the course of this study 725 hand specimens from this limited outcrop have been examined and no definite brachiopod remains found.

DETAILS

The Permian outcrop is confined to the foreshore at Cultra (Figure 7) and its present disposition is largely controlled by faults. It extends for a distance of 320 m W from the fault at the Sewage Works at Cultra, and it rests unconformably upon the Lower Carboniferous shales and cementstones.

The Sewage Works Fault [414 810] trends at 340° and downthrows the Permian rocks against Lower Carboniferous rocks to the east. The Permian rocks dip at 25° towards 315° for a distance of 90 m where they are cut by an olivine-dolerite dyke trending 310° emplaced along the line of a fault which downthrows to the west. This, however, has a displacement of only a few centimetres and beyond it the beds continue to dip at 25° towards 315° until they are intersected by a 1.5 m thick dyke trending at 355°, now entirely converted to rotten lithomarge which, in a short distance, joins into the 310° trending dyke. At the kaolinised dyke there is a swing in strike of the beds which now dip at 25° towards 345° and they continue in this attitude for a further 18 m until they are again cut by a dyke emplaced along a fault. Both fault and dyke trend at 340° and the fault downthrows to the east. West of the fault the beds dip at 10° towards 345° in which attitude they continue until, within 45 m, they pass out to sea. In this part of the section west of the 340° trending fault and passing down the succession towards the Lower Carboniferous the beds are repeated by a fault trending 80° and downthrowing to the south bringing red marls (Permian) to the south against Magnesian Limestone to the north. This fault is truncated eastwards by the 340° trending fault previously described. Passing southwards from this fault is a small syncline, trending at 80°, in red marls (Permian) which broadens their outcrop here to 14 m. Magnesian Limestone, which occurs south of this is anticlinally folded about an axis trending at 80°. This fold is a perfect pericline and within the core a small oval patch of breccia which forms the basal member of the succession is exposed. The Magnesian Limestone extends for 18 m southwards from the rest of the anticline when it passes into 0.6 m of basal breccia which rests unconformably upon the Lower Carboniferous.

A small outcrop of Magnesian Limestone is seen about 1.5 km down the coast towards Belfast, immediately north of Benedar Hockey Ground [404 800]. It is adjacent to a fault and its total area of not more than 3.5 m^2 is strongly sheared; it is not possible to determine strike or dip of the strata. Its presence at this point is due to a fault 650 m further east at Drumlough House which trends at 330° and brings sandstones of the Sherwood Sandstone Group to the west against Lower Carboniferous to the east thus repeating the whole succession and setting the Magnesian Limestone some distance inland, where it is entirely obscured by drift.

EJC

CHAPTER 6

Triassic

INTRODUCTION

The Triassic rocks of the Carrickfergus area are divisible into three lithostratigraphical units; the Sherwood Sandstone Group, formerly known as the Bunter Sandstone; the Mercia Mudstone Group, formerly the Keuper Marl; and the Penarth Group, the former Rhaetic.

The Sherwood Sandstone Group has a very extensive outcrop, although it is mainly beneath the waters of Belfast Lough and only at Cultra and Greencastle are limited exposures to be seen. Most of our knowledge of this group comes from boreholes outside the area of the sheet. Within this district it is about 350 m thick, probably thickening somewhat to the east.

The Mercia Mudstone Group crops out in the low, drift-covered, ground on the north side of the Lough. Although the lower part of the group is exposed on the foreshore and there are limited exposures inland, knowledge of these rocks is mainly from borehole records (see Figure 9). In particular, the saliferous beds are known only from borehole and mine records. Because of the greater thickness of halite beds to the north-east the total thickness of this group increases from about 300 m in the west to probably twice this in the east.

The Penarth Group which is often associated with Lower Lias strata, crops out at the bottom of the scarp formed by the Ulster White Limestone and the basalts and is usually obscured or disturbed by landslips. Thus, the total thickness of some 15 m is deduced from records outside the area of this sheet.

Mostly, the Triassic rocks in this area dip gently to the north-west, although on the foreshore near Greenisland and to the east of Carrickfergus, the Mercia Mudstone Group is disposed in a series of small folds. Elsewhere steep dips, up to 60°, are recorded.

The geophysical evidence in the area, together with the results of the Larne Borehole, indicate that the Triassic rocks lie on the south-east margin of a considerable basin which affects the younger Mesozoic rocks. This sag is centred on the Larne–Glenarm area and, although the top of the Triassic only falls from about 100 m OD in the west to near sea level at Larne, the base of the Triassic sequence falls from approximately −400 m to at least −2000 m OD in the deepest part of the basin.

SHERWOOD SANDSTONE GROUP

The main exposure of this group is on the foreshore near Cultra, on the south side of Belfast Lough. There is a smaller patch on the foreshore near Greenisland, on the north side of the Lough. The beds are dominantly red fine-grained sandstone although there are some yellow or grey bands and occasional red and green mudstones.

Exposures outside the immediate area of the sheet indicate that bands of desiccation breccia, mud-cracks, ripple-marks, rain-pits and similar shallow-water phenomena are common in this group. Some of these features have been noted in the Cultra exposures.

Although the overall dip of the succession is to the north-west, numerous gentle folds occur in the foreshore sections. The estimated thickness is of the order of 350 m.

MERCIA MUDSTONE GROUP

The group consists largely of fine-grained sediments, usually mudstones but with some siltstones and silty mudstones. The predominant colour is red-brown but grey, green and chocolate coloured beds occur, and greenish reduction spots are not uncommon. Sandy bands (skerries) are known to occur; the sandstone beds seen on the foreshore north-east of Carrickfergus Castle represent one of these units. Some of the argillaceous beds are calcareous or dolomitic; gypsum is locally abundant in veins and anhydrite as irregular nodules. Bedded halite—rock salt—occurs in the lower half of the group and is of economic importance.

From an engineering point of view the rocks of the Mercia Mudstone Group are rippable but have a high bearing strength when undisturbed. They weather fairly rapidly but stand up well in properly graded excavations. In steep slopes, however, their stability is suspect and the series of landslips beneath the basalt escarpment at Knockagh are in failed mudstones of this group.

The group has been subdivided into six formations on the basis of the lithology in the deep boreholes at Port More and Larne (Wilson and Manning, 1978; Manning and Wilson, 1975). Apart from the uppermost division, the Collin Glen Formation, these are not readily recognisable in small isolated exposures, and although their thicknesses, and those of the included salt members, are known at Larne, it is improbable that they would be recognisable in the Carrickfergus area due to lithological changes occurring towards the margin of the basin. The succession is:

Collin Glen Formation (Tea Green Marls)
Port More Formation
Knocksoghey Formation
Glenstaghey Formation (including the Larne Halite Member)
Craiganee Formation (including the Carnduff and Ballyboley halites)
Lagavarra Formation

The Collin Glen Formation, distinctive because of its grey-green colour and position immediately below the Penarth Group, is only exposed within this sheet in Woodburn Glen, and at Cloghfin Port where some 2 m of it occurs below shales of the latter group. Though this formation is probably present to the east of Woodburn, the evidence, in landslips below Knockagh, indicates that, like the Penarth Group and the Lias, the upper part of the Mercia Mudstone Group has been removed by pre-Cretaceous erosion in that area. On the foreshore at Lonsdale House and Raven Hill, the red mudstones overlying the Sherwood Sandstone Group are doubtless representative of the Lagavarra Formation, but the remaining exposures cannot be assigned to any specific formation.

The Bedded Halites

The history of salt mining in the area is discussed in Chapter 16. Virtually all the information on the evaporites comes from records, sometimes obscure, of shafts and boreholes sunk during the latter half of the 19th, and first half of the 20th centuries (Figure 9). During the 19th century working of salt took place in four areas: Woodburn (Duncrue), Kilroot, Eden and Red Hall and, at present, rock-salt

is being worked at Kilroot. Apart from the incomplete knowledge of the succession at each of the four former areas the only other information comes from abortive trials at Knockagh Cottage, Clipperstown, Glenview and Eden Cottage, referred to in the history of mining.

No borehole in this sheet has completely penetrated the Mercia Mudstone Group so the total thickness is not known with precision (Figure 10). At Larne, about 7 km N of the sheet boundary, the upper-most salt bed is 254 m below the top of the Mercia Mudstone Group (Figure 11). There, the thickness of this group, less that of the major salt beds, is about 500 m. In the Maiden Mount No. 2 Borehole, located close to the outcrop of the Ulster White Limestone, the top-most salt was 223 m below the base of the drift. The Mercia Mudstone has so far not been bottomed in a borehole in this sheet but, in the shaft and borehole at Clipperstown, the 225 m of mudstone recorded probably lies below the saliferous horizons, giving a minimum aggregate thickness for the Mercia Mudstone Group in this area

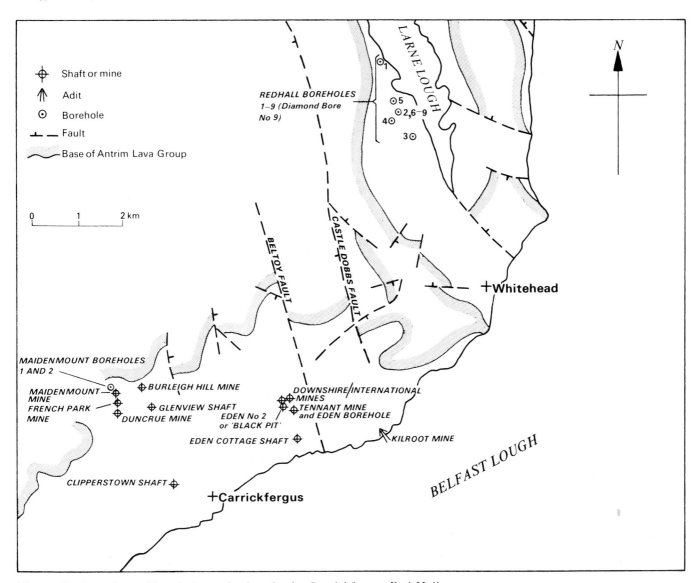

Figure 9 Locations of boreholes and mines in the Carrickfergus–Red Hall area

of at least 535 m.

Apart from the borehole at Larne (Manning and Wilson, 1975), no shafts or bores of which the records are public have completely penetrated the halite horizons. Every gradation between uncontaminated mudstone and pure halite is seen in borehole cores and mine sections. Most of the halite contains enough impurity to give it a reddish brown colour and it is almost invariably banded; commonly with thin layers or films of mudstone concordant with the banding. There are rare pods or veins of clear halite, presumably recrystallised. Small faults and folds in the halite have been noted in mine sections, but some of these are probably due to mining activity and subsequent pressure adjustments.

Where the salt beds outcrop beneath the drift is uncertain and there is no evidence of the wet-rock-head conditions which occur in the Cheshire salt field. There, sub-drift water has dissolved the halite for great distances down-dip and produced highly brecciated and disturbed strata attributed to the collapse of the overlying mudstones. Broken strata are known from boreholes in the Kilroot area but there is some indication that these may be due to Tertiary igneous activity rather than wet-rock-head collapse phenomena. 'Wild brine', produced by groundwater coming into contact with the halite deposits is associated with wet-rock-head conditions in Cheshire. In Antrim there is no evidence of the existence of 'wild brine', although a few salt springs have been recorded. PIM, HEW

In considering the non-occurrence of halite in some boreholes and shafts where it could have been expected, it is interesting to note the variation in amount and direction of dips recorded on some old mine plans. In the International Mine the dip was 14°W, in Tennant 11°SW, while the one dip recorded from the Duncrue salt field, in the French Park Mine, is 8°NE.

When these dips are considered in conjunction with the folding, along predominantly NNW–SSE axes, observed in the Mercia Mudstone Group on the foreshore between Greenisland and Kilroot, it is apparent that this, in conjunction with faulting, accounts for the apparently haphazard distribution of halite in the vicinity of Carrickfergus and that, although the regional dip may be northwards towards Larne, there are substantial corrugations which affect the halite outcrop pattern.

There is also some evidence that the halites are not in beds of constant thickness, although whether this is an original feature or is due to plastic flow associated with folding is not clear. AEG

The information from shafts and boreholes is inadequate both in quantity and detail to allow firm correlations across the area. Nevertheless a tentative correlation is attempted in Figure 10, which implies that halite deposition com-

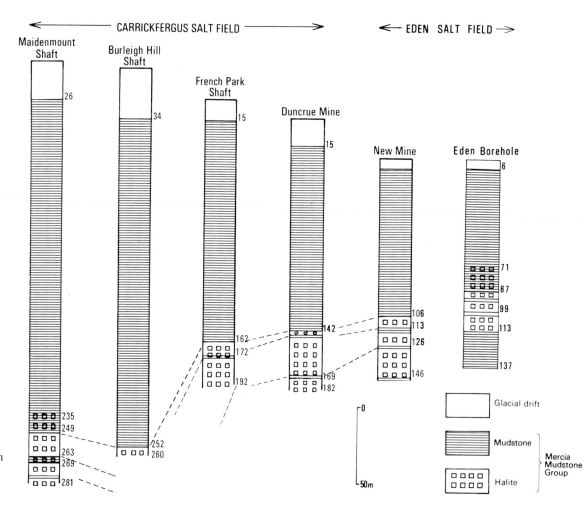

Figure 10 Comparative sections of the Mercia Mudstone Group between Carrickfergus and Eden

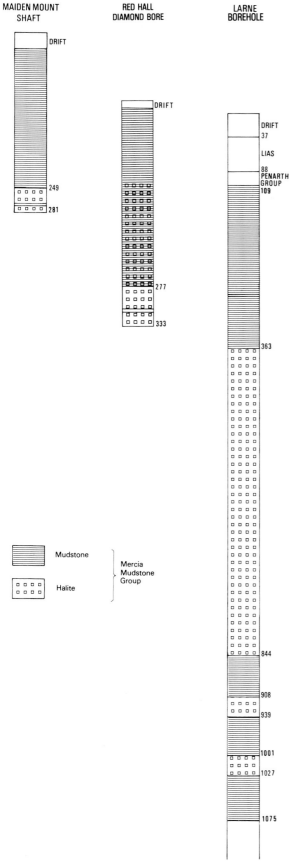

DRIFT

249

281

DRIFT

277

333

DRIFT
37

LIAS

88
PENARTH
GROUP
109

363

844

908
939

1001
1027

1075

▭ Mudstone

▭ Halite

Mercia
Mudstone
Group

Figure 11 Comparative sections of the Mercia Mudstone Group between Carrickfergus and Larne

menced later at Carrickfergus than at Larne, but that the end of the period of halite deposition took place at about the same stratigraphic level throughout the basin—approximately 250 m below the base of the Penarth Group. In addition to the thinning of the salt sequence, due, possibly, to the absence of the lower halite beds around Carrickfergus, the actual thickness of the salt beds is also less because they accumulated closer to the more stable margin of the basin. Similarly, to the west of Duncrue, the halite beds probably die out quite rapidly and salt is not represented in the Mercia Mudstone Group, other than by isolated casts of rock-salt crystals, within the area of the Belfast Sheet (One-inch Sheet 36).

PENARTH GROUP

Towards the close of Triassic times, following the deposition of the Collin Glen Formation, the sea inundated the area and a succession of shales, siltstones and thin limestones was deposited disconformably on the beds of the Mercia Mudstone Group (Figure 13). These deposits, originating in a transgressive marine environment, constitute the Penarth Group, formerly the Rhaetic. The term 'Westbury Formation' has been introduced for the former Lower Rhaetic 'Black Shales' and 'Lilstock Formation' for the Cotham Beds and Langport Beds (Warrington and others, 1980).

The Penarth Group and the overlying Lower Lias exist under much of Antrim but only outcrop around the margins of the lava plateau, where they are much affected by landslips and surface disturbances. Locally, these beds have also been removed by pre-Cretaceous erosion and the very scattered nature of the exposure makes it difficult to determine whether they are present at any particular locality. Within the limits of this sheet, however, the only area where they are undoubtedly missing is in the extreme west, below Knockagh. There, in a temporary pit the Hibernian Greensands Formation rests directly on the Mercia Mudstone Group.

The stratigraphy of the Penarth Group in south-east Antrim has been discussed by Ivimey-Cook (1975) and the exposures at Cloghfin Port and Whitehead railway cutting described. This is the first general reappraisal of the unit in Ireland since the classic papers by Tate (1864, 1867). Tate's work was mainly on the sections at Collin Glen and Waterloo, Larne but he does refer to Woodburn and Whitehead, which are within this area. He followed Moore (1861) in dividing the sequence into an upper part—the 'White Lias', and a lower part—the 'Avicula contorta shales' and, although agreeing with Sir Richard Griffith in regarding these rocks as a sub-formation of the Lias, he pointed out that they were all older than the Lower Lias.

In south-east Antrim, the greenish grey mudstones of the Collin Glen Formation are succeeded, above a sharply defined erosional break, by the lowest dark grey mudstones of the Westbury Formation. This formation is predominantly argillaceous but includes some silty and sandy bands, particularly towards the base, in which the small bivalve *Rhaetavicula contorta* is locally abundant. In the

Larne area it is about 13 m thick, and though no complete section is seen it is estimated that the thickness at Whitehead railway cutting may be up to 10 m. An incomplete section is also seen at Cloghfin Port. The overlying Lilstock Formation is dominantly calcareous mudstones with light coloured silty partings but with some interbedded limestones and darker grey mudstones. It is generally paler in colour than the Westbury Formation. Some of the former show cross-bedding and slump structures; these can be seen in the hornfelsed strata in Woodburn Glen North. In the Larne Borehole, the Lilstock Formation is about 7 m thick but only fragmentary sequences are seen within this sheet, notably in the Whitehead railway cutting and at Cloghfin (Figure 12).

HEW,PIM

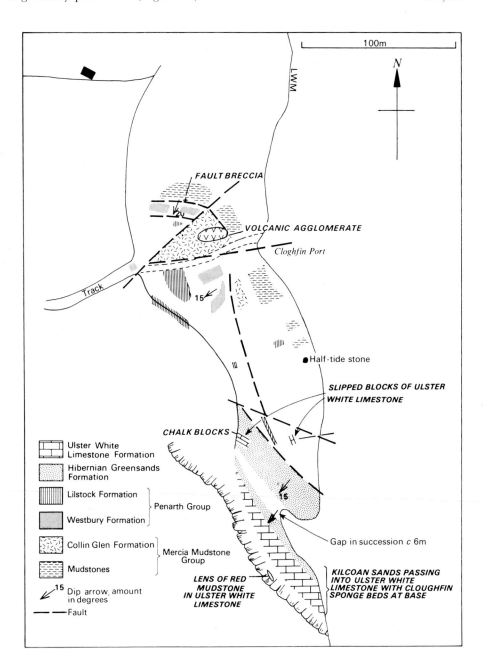

Figure 12 Sketch-map of Cloghfin Port showing the distribution of Triassic and Cretaceous exposures

CHAPTER 7

Jurassic

LOWER LIAS

Like the underlying Penarth Group, upon which they rest with apparent conformity, the soft grey mudstones and shales of the Lias are very rarely exposed within the area of this sheet as the outcrop is greatly affected by landslip and masked by drift deposits. Such exposures as are recorded are often of a temporary nature.

In Knockagh escarpment, near the western edge of the sheet, there is evidence that both the Lias and the Penarth Group, with an unknown proportion of the Mercia Mudstone Group, have been removed by pre-Cretaceous erosion but elsewhere in the district it is assumed, in the absence of contrary evidence, that the Lias is present and the outcrop indicated on the published one-inch sheet is based on this premise. Localities where the Lias has been proved are Lough Mourne; Seamount (near Cloghan Point); Copeland Water and in the Whitehead and Bentra boreholes. In none of these was more than a few metres of Lias present, but in the Larne Borehole, some 6 km N of the sheet margin, the Lias was over 50 m thick; and the thickness at Waterloo, north of Larne, is probably over 100 m. In the Belfast area, to the south-west, the greatest recorded thickness is only 12 m.

Lithologically the Lias consists of pale grey calcareous mudstones with silty laminae, grey siltstones, and, in the lower part, fissile grey shales and thin grey shelly limestone ribs. In adjacent areas palaeontological evidence indicates the occurrence only of Hettangian beds, mainly in the *planorbis* Zone, though at Larne, the *liasicus* and *angulata* zones are also proved and there is some evidence for the lowest beds of the Sinemurian.

DETAILS

In Woodburn Glen exposures are poor and ephemeral and previous reports of the occurrence of Lias shale cannot be substantiated.

On the western side of Lough Mourne, near the spring of Lignaca, Du Noyer mapped 'sparing' appearances of Lias shales. The raising of the water level in the lough, when it was made a reservoir, covered this area, but after the dry year of 1972–73 the water level had fallen by some 5 m in October 1973 and permitted inspection of the ground. The inflow from the spring had cut a channel through the laminated silts which floor the reservoir area and exposed, at a point about 300 m E of Lignaca, grey clay with slabs of thin limestone containing *Cardinia* cf. *hybrida*, *Modiolus?* and *Oxytoma inequivalvis*. Though not *in situ* this material is clearly of very local origin and establishes the presence of beds above the base of the Lias in the area west of the Lough Mourne Fault.

A small exposure of grey mudstone with gryphaeform bivalves possibly indicating the *angulata* or *bucklandi* zones was noted about 200 m S of Seamont [462 903] in a steep slope above the railway (Figure 13). In the same area Tate (1867, p. 299) recorded the occurrence of shaly beds with '*Ammonites planorbis*, *Mytilus minutus* and *Hemipedina*'. This would indicate the *Psiloceras planorbis* Zone of the Hettangian.

About 1.4 km NW of White Head a borehole at Whitehead Waterworks [466 920] penetrated through the Cretaceous into 26 m of 'black and blue laminated shale'. In view of the known thickness of the Penarth Group in this area it is most unlikely that these beds, which were not bottomed, are all attributable to that group and it is a reasonable assumption that they are partly Jurassic in age. In the area lying just to the north-west, around Bentra [460 929], 4.88, 2.85 and 1.02 m of mudstones and limestones were found, beneath the Hibernian Greensands Formation, in three boreholes. These beds are regarded as Lias. HCI-C, HEW

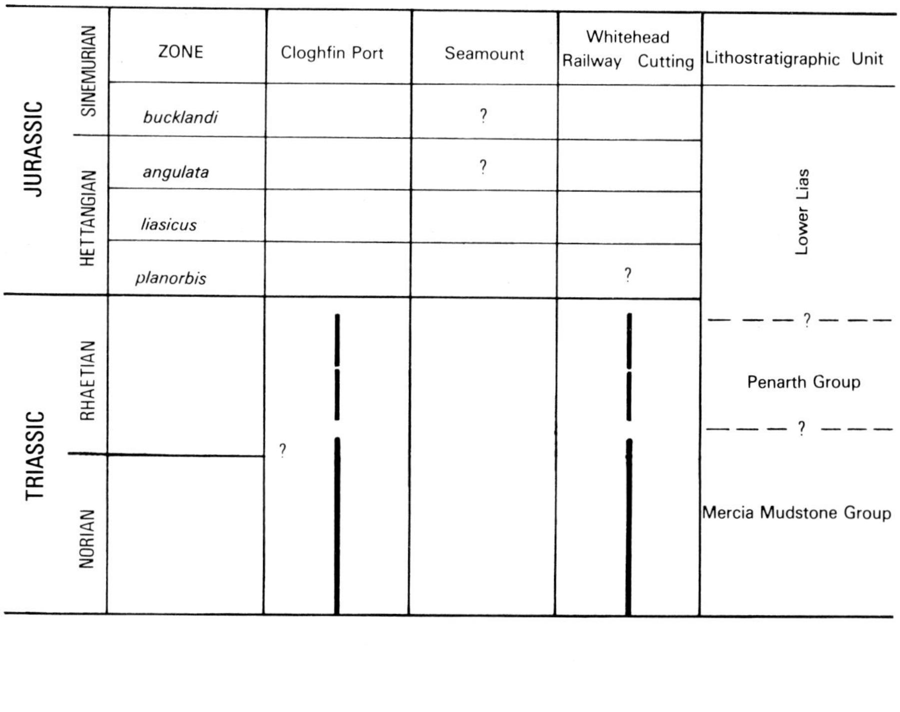

		ZONE	Cloghfin Port	Seamount	Whitehead Railway Cutting	Lithostratigraphic Unit
JURASSIC	SINEMURIAN	*bucklandi*		?		Lower Lias
JURASSIC	HETTANGIAN	*angulata*		?		Lower Lias
JURASSIC	HETTANGIAN	*liasicus*				Lower Lias
JURASSIC	HETTANGIAN	*planorbis*			?	Lower Lias
TRIASSIC	RHAETIAN		?			— — — ? — — — Penarth Group — — — ? — — —
TRIASSIC	NORIAN					Mercia Mudstone Group

Figure 13 Zones and local ranges of the Lias and Penarth Group in the Whitehead area

CHAPTER 8

Cretaceous

INTRODUCTION

In Northern Ireland, the Cretaceous is represented by sediments of Late Cretaceous age, in which a two-fold division can be recognised. The lower subdivision is variably glauconitic, and comprises a sequence of argillaceous and arenaceous lithologies ranging in age from Cenomanian to early Santonian, and collectively known as the Hibernian Greensands Formation (Fletcher, 1978). The upper subdivision consists of chalky limestones with basal glauconite-rich beds, ranging from mid-Santonian to early Maastrichtian, and designated the Ulster White Limestone Formation (Fletcher, 1978).

Throughout this map-area, the Hibernian Greensands Formation varies in thickness from 3 to 20 m, while the Ulster White Limestone Formation has a maximum preserved thickness of 44 m. The relationships between the component members of each formation are shown diagrammatically on Figures 14 and 15, where several breaks in sedimentation are indicated. Locally, members or parts of members may be attenuated or absent due to variations in the amounts of sedimentation and degree of erosion.

DISTRIBUTION AND STRUCTURAL SETTING

Cretaceous rocks in this district are confined to the area north of Belfast Lough. Although these sediments occur over a considerable area, they are largely concealed by basalts of the Antrim Lava Group (Old, 1975), so that their outcrops are restricted to narrow bands along the scarps of the Antrim Plateau. Between Knockagh and Woodburn the outcrop parallels the Caledonoid structural grain of the region, but north-east of Woodburn it is offset by several relatively closely spaced NNW-trending faults. These faults produce a number of small ribbon-like inliers of Cretaceous and pre-Cretaceous rocks within the basalt plateau, and progressively throw down to the north-east, bringing the Cretaceous outcrop down to sea level at Whitehead. Another fault of the same group (the Larne Lough Fault) repeats the Cretaceous outcrop to the east of the Lough on the Island Magee peninsula.

The Carrickfergus district is situated near the southern margin of the east Antrim depositional basin (Fletcher, 1978), which, in gross regional structural terms, can be regarded as the south-westward extension of the 'Midland Valley of Scotland' (George, 1967). This basin is bounded to the south by the Down–Longford Block, corresponding to the 'Southern Uplands'. There is a regional attenuation of sediment towards the block, and the lower members of the White Limestone thin out altogether and are progressively overlapped by higher members in this direction. In addition, Cretaceous sedimentation was strongly influenced by a NNE–SSW-trending structural high called the Knockagh Axis (Fletcher, 1978) extending from Portmuck and Magheramorne in Sheet 21 through Glenoe and Beltoy to Woodburn and Knockagh, and thus striking at an angle to the Caledonoid axis of the 'Midland Valley'. At Knockagh, Cretaceous rocks rest directly on red and green Triassic rocks, rather than on grey mudstones of the Penarth Group and Lower Lias, as elsewhere in the sheet-area. Between Woodburn and Knockagh the various members of the Hibernian Greensands progressively thin on to the axis, and at Knockagh the White Limestone almost oversteps on to the basal member (Belfast Marls). The Knockagh Axis delimits a sedimentary basin to the south-east, known as the Hillsport Basin (Reid, 1971). Here the highest member of the Hibernian Greensands (Kilcoan Sands) is most completely preserved, and the Cloghfin Sponge Beds Member at the base of the White Limestone succession also reaches its maximum development.

Figure 15 (extensively modified from Fletcher, 1978, fig. 8) shows the effect of structural control of Cretaceous sedimentation within the Carrickfergus district and in the adjacent map-areas. Within the 'Midland Valley' the preserved sequence of the White Limestone does not extend higher than a level in the Garron Chalk Member. This situation contrasts with that obtaining in the North Antrim Basin (Sheets 7 and 8), and on the Down–Longford Block, where the succession extends up into the Port Calliagh Chalk Member.

HIBERNIAN GREENSANDS FORMATION

This formation comprises four distinct clastic sedimentary units for which a formal lithostratigraphical nomenclature has been erected (Fletcher and Wood, in press). The new nomenclature has been adopted in this account.

All the sediments of the Hibernian Greensands Formation were deposited before *Uintacrinus socialis* Zone time, when limestone deposition became established. There is evidence, however, that the main period of limestone deposition was preceded by shorter episodes, but that the chalk beds deposited then were removed by intraformational erosion, to survive only as relict infillings of larger fossils in the upper beds of the Hibernian Greensands.

The lowest three members are of Cenomanian–Turonian age, and are separated from the upper Coniacian–Santonian member by a non-sequence which has given rise to a pronounced unconformity.

Cenomanian–Turonian greensands

BELFAST MARLS MEMBER

The basal unit of the Cretaceous throughout the map-area comprises a thin sequence of dark green glauconitic marls,

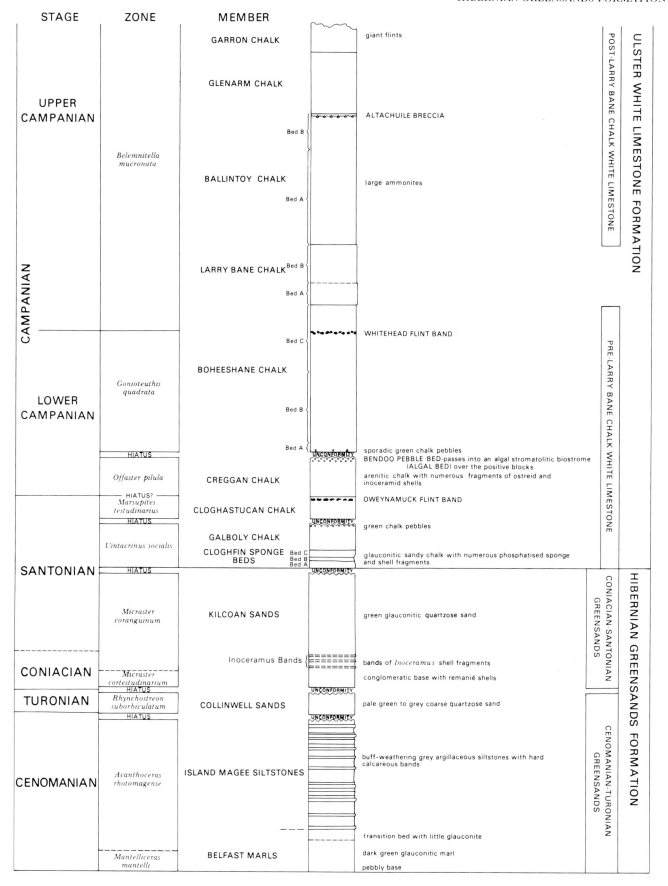

Figure 14 Cretaceous stratigraphy

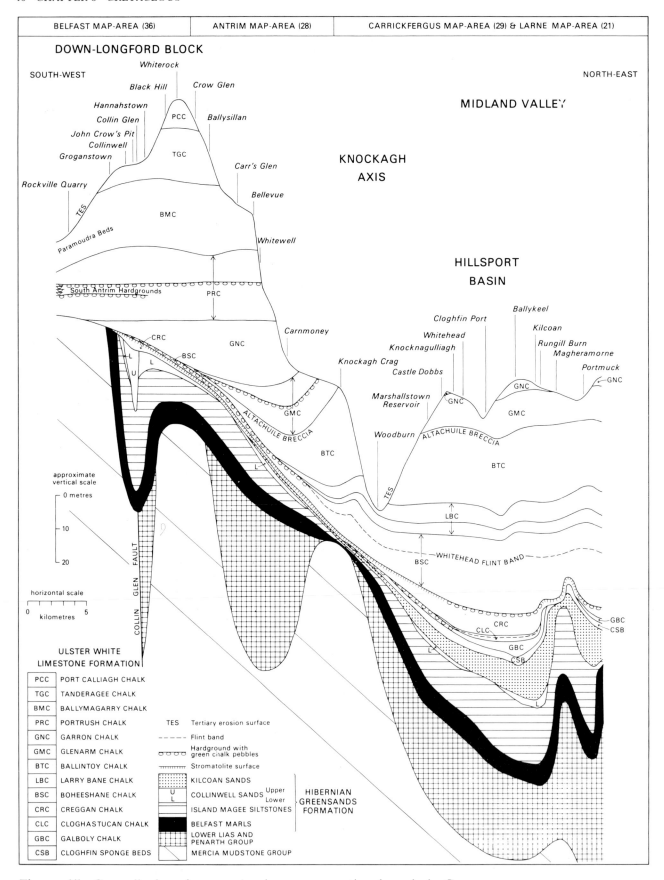

Figure 15 Generalised south-west to north-east cross-section through the Cretaceous sequence of south-east Antrim, illustrating lateral variation in thickness of constituent members

formerly known as the Glauconitic Sands (Tate, 1865). The transgressive base of this member oversteps Lower Lias mudstones at Whitehead, and the Penarth Group near Woodburn, to rest on Mercia Mudstones along the Knockagh escarpment. Contact with the overlying member is conformable, and the junction is arbitrarily drawn at the base of the lowest persistent hard calcareous siltstone at the top of a transitional zone of decreasing glauconite content (Figure 14). Contrary to the lithology suggested by the earlier name, this member comprises pale grey strongly bioturbated silty marls, with a scattering of dark green glauconite grains, not exceeding 40 per cent. All the beds are fossiliferous, but the fossils tend to be concentrated into discrete and laterally continuous shell-beds.

The three known exposures of this member in the district occur at Cloghfin Port, where a 5-m-wide outcrop of gently dipping strata is indifferently exposed at low tide, at Woodburn Glen South, and in a small quarry within the slipped ground 503 m north-west of Knock Lodge. The thickness is thought to be a little less than the 3 m recorded in a borehole near Magheramorne (Sheet 21), reducing progressively to a minimum of 1.6 m over the Knockagh Axis.

A shell bed in the section at Carrs Glen (Sheet 28) yields an assemblage of poorly preserved moulds of *Schloenbachia* and subordinate *Mantelliceras* indicating an early Cenomanian (*Mantelliceras mantelli* Zone) age. However, the occurrence of *Acanthoceras* at Cloghfin, and the belemnite *Actinocamax primus* at Cloghfin, Woodburn, and in the former exposure at Whitehead indicates that the upper part of the Belfast Marls may be of mid-Cenomanian age.

The Belfast Marls yield a diverse fauna which is dominated by bivalves, but also includes serpulids, brachiopods, gastropods, cephalopods and selachian teeth. No echinoderms have been recorded. The large exogyrine oyster *Amphidonte* aff. *obliquata* and an undescribed species of *Entolium*, known also from early Cenomanian greensands in southern England, Germany and the USSR, are particularly common. Many fossils were found during the excavations for the railway tunnel near Whitehead (Tate, 1865), and also at Woodburn Glen, but today the best collecting locality in the sheet-area is Cloghfin. The latter locality is remarkably rich in small brachiopods and pectinids, a fact noted by Hume (1897). The bivalve fauna is intermediate in character between that of the late Albian and early Cenomanian greensands of southern England and that of the (Cenomanian) Lower Chalk. Species collected in the Carrickfergus district include the following in addition to the ubiquitous *Amphidonte* aff. *obliquata* and *Entolium sp. nov.*:

Acesta sp., Camptonectes cf. *virgatus, 'Chlamys'* aff. *fissicosta, 'C.'* cf. *hispida, Entolium balticum, Gryphaeostrea canaliculata, Merklinia aspera, Nanonavis sp., Neithea (Neithella) notabilis, N. quinquecostata* s.l., *N. sexcostata, Ostrea* aff. *walkeri, Oxytoma seminudum, 'Pecten' glauconeus* ['*Aequipecten' arlesiensis*?], *Plicatula deltoidea, Rastellum sp., Rhynchostreon plicatulum* and *Thetis sp.*

The transitional zone with relatively little glauconite at the top of this member was treated as a separate entity by Hume (1897), and named the 'Glauconitic Marls'. This part of the sequence is lithologically part of the Belfast Marls Member, but yields a fauna which is quite distinct from that of the earlier strata, and which is linked more closely with that of the overlying Island Magee Siltstones Member. The thick-shelled *Amphidonte* aff. *obliquata* has not been found, and the assemblage is dominated by *Gryphaeostrea canaliculata*, with subordinate '*Chlamys*' aff. *fissicosta* [almost smooth, in contrast to the broadly ribbed forms in the lower beds], *Entolium* cf. *laminosum, Neithea sp.* and '*Inoceramus*' ex gr. *reachensis*. The serpulid *Rotularia rotula* was collected at Woodburn.

ISLAND MAGEE SILTSTONES MEMBER

This member comprises an alternation of hard and soft grey siltstones and marls that weather to a pale yellow colour, giving rise to the earlier name of 'Yellow Sandstones and Grey Marls'. It conformably succeeds the underlying Belfast Marls, but is disconformably overlain by younger strata. When fresh, the sediments are pale blue-grey, and distinctly calcareous, but on exposure they undergo slight decalcification. Very little glauconite or sand is present, and the top surface of the member is usually bored, with pipes of younger sediment extending down for over 0.6 m into the siltstones, eg at Woodburn Glen South.

Detailed measurement of the available exposures indicates that the banding is stratigraphically controlled, and that individual bands can be traced over considerable distances. The Island Magee Siltstones are best developed in Sheet 21, where they have a maximum thickness of 8 m in the Magheramorne area. In the Carrickfergus district the most complete sequence preserved is at Cloghfin, where at least 7 m are exposed as foreshore reefs. Sedimentary thinning and differential erosion over the Knockagh Axis have resulted in a reduction to a mere 0.67 m at Woodburn.

The member can be assigned a mid-Cenomanian age on the occurrence of *Euomphaloceras cunningtoni* and other acanthoceratid ammonites (Hancock, 1961; Kennedy, personal communication), the assemblage probably equating with the *Turrilites acutus* assemblage of the *Acanthoceras rhotomagense* Zone. Also of biostratigraphical significance is the inoceramid '*Inoceramus*' *hamiltoni* [= '*I.*' *atlanticus*?], which is closely related to '*I.*' *reachensis* from the higher part of the mid-Cenomanian Lower Chalk succession of southern England. The occurrence of one or more species of the irregular echinoid '*Epiaster*' in the Island Magee Siltstones suggests the presence of southern ('Tethyan') elements in the fauna.

Hume (1897) pointed out that the fauna of the Island Magee Siltstones on the east coast was much less diverse than that of the sections near Belfast, such as Collin Glen, and noted the relative abundance of small corals and the echinoid '*Discoidea subuculus*'. No echinoids agreeing with this description were found during the resurvey, but the small corals *Micrabacia* cf. *coronula* and '*Stephanophyllia*' cf. *bowerbankii* were collected from several horizons at Cloghfin. A comparable occurrence is found at an approximately equivalent horizon in the Lower Chalk of southern England. In addition to the corals, the following fauna was collected from Cloghfin during the resurvey:

Rotularia spp. [including carinate forms], *Entolium* cf. *laminosum, Gryphaeostrea canaliculata, 'Inoceramus'* cf. *hamiltoni, Neithea aequicostata, N. quinquecostata* s.l., *Pseudoptera sp., 'Epiaster' sp.* and *Holaster* cf. *trecensis*. The large serpulid *Hepteris septemsulcatum* occurs at Whitehead.

COLLINWELL SANDS MEMBER

In many areas of south and east Antrim, a variable thickness of coarse sands, formerly known as the Lower Glauconitic Sandstone, intervenes between the Island Magee Member and the overlying (Coniacian–Santonian) Kilcoan Sands Member. This sequence is best developed south-west of Belfast (Sheet 36), where the youngest beds, comprising unfossiliferous white quartzose sands, are preserved. Typically, the member is characterised by pale green glauconitic coarse-grained unconsolidated sands with the large exogyrine oyster *Rhynchostreon suborbiculatum* [formerly known as *Exogyra columba*]. In the Carrickfergus district, the Collinwell Sands comprise poorly glauconitic whitish sands resembling the quartzose sands near Belfast, but the occurrence of *R. suborbiculatum* suggests that only the lower part of the member is represented. A maximum thickness of over 2.4 m is preserved at Rungill Burn.

The dating of the Collinwell Sands member is somewhat uncertain. The member was formerly assigned a Cenomanian age (Hancock, 1961) on the basis of the occurrence of *R. suborbiculatum*, but this oyster is known to range from late Cenomanian to mid-Turonian in many parts of Europe. Two unlocalised *Inoceramus* ex gr. *lamarcki* (Ulster Museum) from this member are of mid-Turonian aspect.

Hume (1897) recorded a glauconitic sandstone 'crowded with *Rhynchonella schloenbachi*' from 'south of the tunnel, near Whitehead'. A block in the Ulster Museum collected in 1895 and agreeing with this description contains numerous rhynchonellids of uncertain generic affinity associated with *Hepteris septemsulcatum*, '*Stephanophyllia*' cf. *bowerbankii* and a fragment of a large *Neithea*. The lithology matches material from the lower part of the member in Collin Glen (Sheet 36).

Mid-Hibernian greensands unconformity

A non-sequence covering much of the Turonian and probably some of the Coniacian can be recognised within the greensands. A conglomerate of varying coarseness rests on eroded remnants of the lower three members and, in the more positive areas, rests directly on an exhumed pre-Cretaceous floor. Within the Carrickfergus area, the oldest strata in direct contact with the conglomerate belong to the Island Magee Siltstones Member.

Coniacian-Santonian greensands

KILCOAN SANDS MEMBER

The above-mentioned conglomerate forms the basal component of a sequence of pale green glauconitic sands which largely results from the erosion and redeposition of the sands of the Collinwell Member. The type-section is exposed at Kilcoan on Island Magee (Sheet 21), and correlations based on this succession show that considerable erosion took place prior to the main period of chalk sedimentation (see Figure 16).

The member is usually overlain unconformably by the Cloghfin Sponge Beds (Plate 13), but, towards the positive areas, younger chalk members overstep one another on to the greensands. However, throughout this map-area, the greensands are immediately overlain by the Cloghfin Sponge Beds, although over the Knockagh Axis at Woodburn the

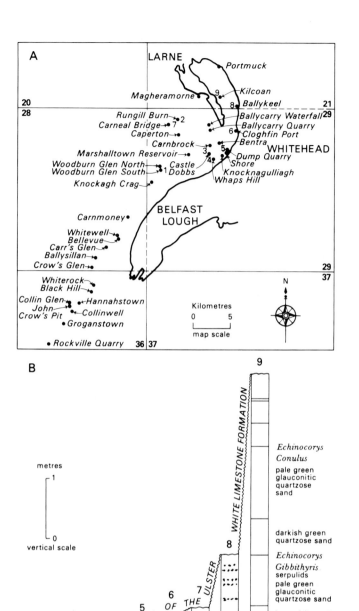

Figure 16 A. Location map of the main Cretaceous exposures in south-east Antrim
B. Comparative vertical sections in the Kilcoan sands

greater part of the Kilcoan Sands has been removed by erosion.

Within the Kilcoan Sands Member, three conspicuous shell bands packed with *Inoceramus* fragments aligned more or less parallel to the bedding (the Inoceramus Bands) allow the succession to be divided into three unequal parts. The pre-Inoceramus Bands succession is characterised by *Conulus raulini* and a diverse assemblage of rhynchonellids including *Cretirhynchia* aff. *octoplicata*, '*C.*' *robusta* [originally described from Woodburn] and *Cyclothyris sp*. The association of *Conulus raulini* and '*Cretirhynchia*' *robusta* suggests an early Coniacian age. The Inoceramus Bands can be dated on inoceramids, since in addition to fragments of *Inoceramus* '*cuvieri*', they yield pieces of the distinctive cap-valves of *Volviceramus involutus,* a Coniacian zonal index fossil which occurs near the base of the *Micraster coranguinum* Zone of the southern England Chalk and correlative deposits elsewhere in Europe and North America. The Inoceramus Bands also yield numerous oysters including *Gryphaeostrea canaliculata*, large plicate *Hyotissa semiplana*, and a small globose pycnodonteine which may, as at Ballycarry Waterfall section,

occur in clumps of juvenile individuals. Associated with the oysters are *Merklinia sp.*, *Neithea quinquecostata* s.l., *Spondylus spinosus*, *S. sp*. and an assemblage of terebratulids including species of *Gibbithyris* and *Concinnithyris?*

The post-Inoceramus Bands succession contains numerous specimens of the large flat terebratulid *Gibbithyris hibernica*, especially at the base. In condensed successions along the Knockagh Axis, eg Woodburn, this brachiopod occurs *within* the Inoceramus Bands, but towards the basin, eg Castle Dobbs, Whaps Hill, the first definite records appear to be from above this horizon. Higher in the succession small and medium-sized *Conulus sp*. occur associated with rare depressed *Micraster sp*. and common *Echinocorys sp.*, and probably denote the beginning of the Santonian. To judge from the matrix, a magnificent crushing tooth of the ray *Ptychodus belluccii* from the Ballycarry Waterfall (Baily *in* Hull, 1876) is from this part of the succession. One or more horizons characterised by clumps of the serpulid *Sarcinella socialis* [*Serpula filiformis* of earlier literature] occur near the middle of the fully-preserved post-Inoceramus Bands succession.

Plate 13 Hibernian Greensands and basal Ulster White Limestone. Kilcoan Sands Member overlain by cobbly-weathering chalk at top. Cliff south of Cloghfin Port   [NI 358]

ULSTER WHITE LIMESTONE FORMATION

The strata assigned to this formation include all the whitish-coloured chalky limestone lying above the Hibernian Greensands Formation and below the Tertiary 'Clay-with-Flints' deposit. They range in age from Santonian (*Uintacrinus socialis* Zone) to early Maastrichtian (*Belemnella occidentalis* Zone), and the formation is disconformable at its lower and upper limits. The formation is subdivided into 14 members (Fletcher, 1978) which appear to maintain their identity over considerable distances. Details of the biostratigraphy of each member have been given for the North Antrim basin successions (Fletcher and Wood, 1978), and the limited collecting carried out in the Carrickfergus map-area suggest that there are no significant differences, except where condensation has occurred over the Knockagh Axis. For this reason, only the Cloghfin Sponge Beds will be considered in detail in this account.

The interpretation of the junction between the greensands and the chalk is controversial, and two main views are held. The first considers that there was a gradual change from greensand to chalk deposition, and that this change was diachronous across the margins of the subsiding chalk basin. The alternative, and the one favoured here, recognises a differential non-sequence between two chronologically separate episodes of deposition, and considers that the variable unit of glauconite and sand-bearing limestone about the junction largely represents contamination of chalk sediment from the pre-existing subjacent greensands. Thus the diachronous base of the limestone reflects a changing submarine front of chalk sedimentation with no associated transgressing sandy shoreline (Figure 15).

Several periods of non-deposition can be recognised in the chalk sequence, indicated by minor glauconitised erosion surfaces with associated glauconitised chalk pebbles in the basinal facies, and by algal stromatolite-encrusted hardgrounds over the positive areas.

Two prominent chalk posts with a thickness ratio of 1:2 comprise the Larry Bane Chalk Member. This unit is easily recognisable in most sections in east Antrim and elsewhere, and the Ulster White Limestone Formation may be conveniently divided into pre-Larry Bane, Larry Bane, and post-Larry Bane successions (see Fletcher, 1978, p. 7; and Figure 14). Representatives of the lowest nine chalk members are present within the map-area and the preserved thickness of the formation varies from a maximum of 44 m near Ballycarry to about 10 m over the Knockagh Axis at Woodburn.

Pre-Larry Bane Chalk

The transgressive base of the Ulster White Limestone Formation is not easy to demonstrate within the map-area since the Cloghfin Sponge Beds everywhere cover the greensands. Pre-Larry Bane Chalk is present throughout the outcrop, but the Cloghastucan and Galboly Chalks are absent over the Knockagh Axis, where the Creggan and Boheeshane Chalks are generally thin.

Cloghfin Sponge Beds Member

The oldest chalk sediments are highly contaminated with glauconite and are characteristically rich in phosphatised pseudomorphs of hexactinellid sponges. This member is best developed on Island Magee in the Hillsport Basin, but the type-section is taken at Whitehead, on the foreshore close to the railway bridge [476 911], where it attains its greatest thickness of 1.73 m. The Cloghfin Member thins on to the Knockagh Axis, and dies out as an identifiable unit just south of Woodburn (Figure 15).

In the thickest sequences, three separate beds (coded A–C in ascending order) can be recognised (Figure 17). The basal bed takes on a distinctive cobbly appearance on weathering, and it can be conveniently referred to as the 'cobble bed'. Bed B consists of more evenly bedded and slightly less glauconitic chalk with a high proportion of orange-brown pseudomorphs, these being particularly common at the base. Bed C is also rich in sponges, but is only sparsely glauconitic, and much closer to normal White Limestone lithology than to a friable chalk. The marked wavy bedding tends to be accentuated by weathering giving this bed a very distinctive appearance. Bed C terminates in a bedding plane and is overlain by a thin (0.18 m) unit of parallel-bedded limestone containing pale yellow sponges in pebble preservation: this unit forms the base of the Galboly Chalk Member. The relationship between the present and previous classifications of the beds about the formational junction is complicated, and can best be explained by means of a diagram (Figure 17).

The famous sponge fauna comprises almost exclusively hexactinellids, of which typical species are *Camerospongia fungiformis*, *Etheridgia mirabilis*, *Eurete* cf. *formosum*, *Leptophragma striatopunctata*, *Rhizopoterion spp.* and *Wollemannia laevis;* the only lithistid, *Callopegma obconicum*, is rare. The sponges show much evidence of reworking and appear to be derived. Reid (1958) recognised two faunal divisions based on the common species of *Rhizopoterion* present: a lower division characterised by *R.* cf. *tubiforme*; and an upper division with *R. cribrosum* in which the sponges are generally more complete and less strongly phosphatised. The boundary between these two faunal divisions appears to fall within Bed C (Reid, personnal communication). All three beds of the Cloghfin Member yield the distinctive flat-topped variants of *Echinocorys* which characterise the *Uintacrinus socialis* Zone in southern England, and several specimens of the high-conical *Conulus albogalerus* are known from the cobble bed. *Actinocamax verus* enters at the base of Bed B, and is not uncommon: this belemnite is particularly characteristic of the *socialis* Zone of southern England, and its occurrence in the Cloghfin Sponge Beds together with the flat-topped *Echinocorys* appears to support a *socialis* age for the whole member, with the sponges having been reworked from sediments of latest *coranguinum* age. No definite specimens of the zonal index fossil have been found to date, but worn and indeterminate crinoid brachials are known from Bed C.

Galboly Chalk Member

Galboly Chalk comprises sediment deposited in *socialis* Zone time. Throughout the map-area it rests upon the Cloghfin Member and, in deeper parts of the depositional basin, is overlain by the Cloghastucan Chalk Member. Both contacts are erosive, and marked by horizons of glauconitised chalk pebbles. The uppermost 0.45 m are

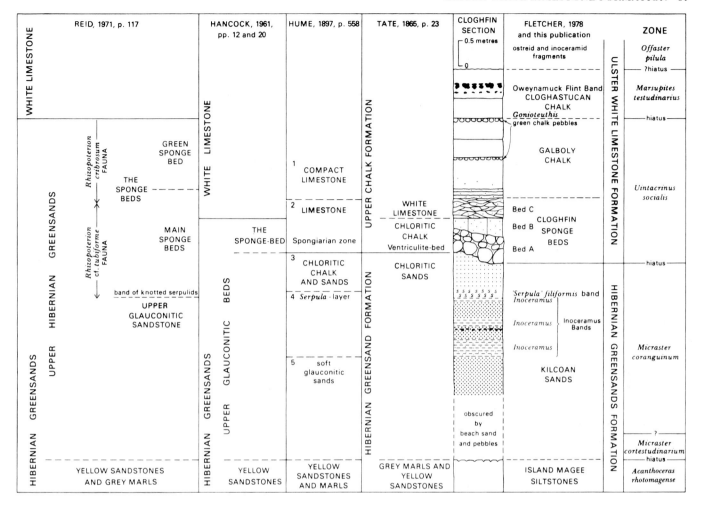

Figure 17 Comparative lithostratigraphical classifications of the Hibernian Greensands–Ulster White Limestone junction exposed at Cloghfin Port

marked by a concentration of green (glauconitised) sponges of the *Rhizopoterion cribrosum* fauna preserved as pebbles. The Galboly Chalk is exposed at Cloghfin, Knocknagulliagh, Castle Dobbs and on the Whitehead foreshore where it reaches a maximum thickness for the area of 2.03 m. Galboly Chalk is flintless at all these localities, but immediately to the north, at Ballykeel, it is flinty and has yielded *Uintacrinus*.

CLOGHASTUCAN CHALK MEMBER

This member is characterised by the occurrence of calyx plates and brachials of the crinoid *Marsupites testudinarius*. In the thickest sequences, a conspicuous band of large closely-spaced flints known as the Oweynamuck Flint Band is developed near the top, and this forms a useful marker. However, in the reduced sequences such as occur within the Carrickfergus map-area, the Oweynamuck Flint Band is generally poorly developed or even absent; where it is absent altogether, recognition of Cloghastucan Chalk in pre-Creggan sequences is based on the occurrence of the belemnite *Gonioteuthis*, which makes its first appearance in this member. Thin representatives can be examined at the

same localities as noted above for Galboly Chalk. At Cloghfin, *Gonioteuthis granulata* and ornamented calyx plates of *Marsupites* can be collected from the thin unit between the green pebble sponge concentrate at the top of the Galboly Chalk and the Oweynamuck Flint Band, the plates being particularly common on the well marked bedding plane just below the flint band; *Marsupites* plates also occur at Knocknagulliagh.

CREGGAN CHALK MEMBER

The chalk sediment of the Creggan Member is coarsely arenitic with a high content of comminuted *Inoceramus* shell. The distinctly gritty texture contrasts with the relatively smooth limestones of the underlying and overlying members. The section at Whitehead is 2.75 m thick, and represents the maximum development of this member in the Carrickfergus district.

The transgressive nature of the base of this member is well demonstrated in this area; it is attenuated towards the Knockagh Axis, and oversteps the Cloghastucan and Galboly members to rest directly on the Cloghfin Sponge Beds at Woodburn (Figures 15 and 18). Two contrasting sedimentary regimes are represented within the chalk of the

Creggan member, one associated with the deeper parts of the basins, the other characterising marginal areas.

1 Where Creggan Chalk immediately overlies non-glauconitic chalk sediment, several horizons of glauconitised chalk pebbles are developed in its upper levels reflecting minor periods of non-deposition and erosion. Similar horizons also occur in the lower part of the overlying Boheeshane Chalk Member, so that the junction between the two members is not always immediately apparent, but there is always a marked change from the gritty Creggan Chalk to the finer-grained limestones of the Boheeshane Member. The interpretation of the Ballycarry Waterfall section has to be based on this criterion.

2 Near the margins of the depositional basins, Creggan Chalk is usually less than 2 m thick, and its gritty texture is much more striking. The top of the member is marked by a stromatolitic biostrome ('Algal Bed'), which appears to represent a coalescing of all the uppermost Creggan non-sequences of the basinal sections. In such areas, the junction between Creggan Chalk and any succeeding member is very clearly differentiated.

Echinoid evidence from outside the map limits permits correlation of the Creggan Member with the *Echinocorys depressula* Subzone of the southern English *Offaster pilula* Zone. The lower beds yield abundant thin shelled oysters identifiable as *Pseudoperna boucheroni*, a species which characterises both the crinoid and lower part of the *pilula* Zones in the English Chalk facies. Examples have been collected at Woodburn Glen North, together with a hinge of the inoceramid *Sphenoceramus*. *Leptophragma* cf. *striatopunctata* was collected from an unspecified horizon in Woodburn Glen North, and the zonal index fossil was found immediately below the Algal Bed. *Gonioteuthis granulata* is present at most localities, and can be collected at Knocknagulliagh.

BOHEESHANE CHALK MEMBER

The highest member of the pre-Larry Bane Chalk is named Boheeshane Chalk. At its type-locality in north Antrim, three distinct beds (A–C) can be recognised (Fletcher, 1978, p. 23). The member has a maximum thickness of 13 m at coastal localities in the Carrickfergus district, attenuating to 6.2 m at Woodburn, where only Bed C is complete. Over the Knockagh Axis there is a more pronounced reduction, and beyond Carnmoney (Sheet 28), the member is only very thinly represented. Of the three beds recognisable at the type-section, only the uppermost (Bed C) retains its identity in this area; Bed A is barely preserved and Bed B is greatly attenuated. The contact with the underlying Creggan Chalk is markedly unconformable.

The pre-Bed C succession is exposed in the coastal sections, and can be readily examined at Whitehead and Cloghfin. The basal metre is flintless, but includes several horizons of glauconitised chalk pebbles. In the type-area, the flinty part of Bed A (A2) is characterised by horizons with *Belemnitella* ex gr. *praecursor* (see Fletcher and Wood, 1978). In the Carrickfergus district, the occurrence of this belemnite in the lower part of the pre-Bed C succession at Ballycarry Waterfall and on Island Magee (Figure 18) provides palaeontological evidence of Bed A. Similarly, the occurrence of the small rhynchonellid *Orbirhynchia bella* about 1 m above the base of the member at Knocknagulliagh may represent the well-marked Orbirhynchia

Band which occurs near the base of Bed B in the type-area.

Bed C contains five separate bands of large flints (see Fletcher and Wood, 1978, fig. 18 and p. 95), and is delineated at the base by a well marked bedding plane. The middle of the five bands marks the highest known occurrence of *Gonioteuthis* in the Northern Ireland succession, and thus constitutes the local boundary between the zones of *Gonioteuthis quadrata* and *Belemnitella mucronata*, corresponding to the boundary between the Lower and Upper Campanian Substages. This conspicuous and ubiquitous marker is named the Whitehead Flint Band after the old quarry [474 913] now used as the Whitehead Dump (see Fletcher, 1978, pl. 4.3).

Although the Boheeshane Chalk succession is poorly exposed in Woodburn Glen North, it is of considerable interest because of the presence of remanié faunas in pebble beds occurring 0.20 and 0.54 m above the Algal Bed at the top of the Creggan Member. The lower pebble bed yields well preserved coarsely granulate large *Gonioteuthis sp.*, and rolled steinkerns of *Echinocorys* ex gr. *conica*. The higher pebble bed yields indigenous small weakly granulate *Gonioteuthis* associated with pycnodonteine oysters and a pebble preservation assemblage comprising *E.* ex gr. *conica*, small *Micraster sp.*, corroded and bored *Gonioteuthis* and indeterminate (pachydiscid?) ammonites. Indigenous *Cretirhynchia* ex gr. *norvicensis* occur in the chalk between the two pebble beds and the matrix of the higher pebble bed yielded a possible 'cyclothyrid *sp. nov*' (Fletcher and Wood, 1978, fig. 23). The flintless chalk overlying the higher pebble bed is characterised by numerous large *Belemnitella* ex gr. *praecursor*. The rest of the Boheeshane succession here includes several bands of large flints of Bed C type, and terminates at the dirty bedding-plane marking the base of Larry Bane Chalk. In Woodburn Glen South, the Boheeshane succession is even more reduced, and examples of the large *Belemnitella* found above the upper pebble bed in Woodburn Glen North occur only 0.06 m above the Algal Bed, below a thin poorly flinty unit which has yielded an undoubted *Gonioteuthis*.

It is difficult to interpret these reduced successions in the two Woodburn glens with certainty, but it would appear that Bed A and the greater part of Bed B are missing at the levels of pebble concentrations, since *Echinocorys* ex gr. *conica* does not appear in the standard sequence until the upper part of Boheeshane B.

LARRY BANE CHALK MEMBER

This member is present throughout the map-area, and is almost 5 m thick. It is completely exposed at Cloghfin, in the Whitehead Dump quarry, on the Whitehead foreshore and in the large Ballycarry Quarry. In the Alt-fraechan Gorge north-west of Red Hall, these beds are covered with vegetation, while a similar section in Woodburn Glen North is less complete, because Tertiary erosion has cut down into the middle part of Bed B. The highest beds of Bed B are, however, exposed in the Marshallstown Reservoir Quarry.

The characteristic fauna comprising *Cretirhynchia woodwardi*, *Kingena pentangulata*, *Belemnitella mucronata senior*, *Echinocorys* ex gr. *conica* and *E.* ex gr. *marginata-subglobosa* is well represented at the better exposed sections, although collecting conditions are not as favourable as in other areas. A

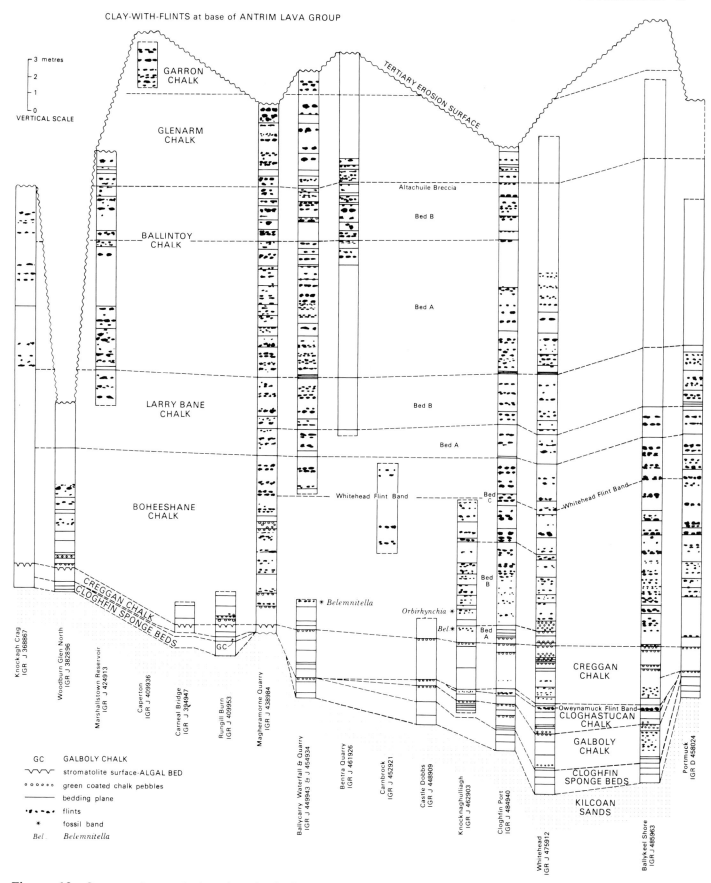

Figure 18 Comparative vertical sections in the Ulster White Limestone

large *Micraster glyphus* and several large *E.* ex gr. *conica* were collected from an echinoid horizon low in Larry Bane B on the Whitehead foreshore.

POST-LARRY BANE CHALK

The remaining thickness of the Ulster White Limestone Formation between the Larry Bane Chalk Member and the Tertiary 'Clay-with-Flints' varies considerably between localities. In the 'Midland Valley' the Tertiary erosional level reached a much lower horizon in the Chalk succession than in other tectonic regions (Fletcher, 1978, fig. 8). This contrast is particularly noticeable when compared with the succession on the adjacent Down–Longford Block, where it comprises younger strata not represented in east Antrim (Figure 15). Within this map-area, the effects of the Knock-agh Axis on the intensity of Tertiary erosion is indicated by the thinner remnants across the Woodburn region, where no post-Larry Bane Chalk is preserved.

BALLINTOY CHALK MEMBER

The top of this member is marked by a conspicuous separation-plane which immediately overlies an irregular band of wavy-bedded chalk containing indurated chalk clasts and angular flint shards, and known as the Altachuile Breccia. At the type-locality, the member can be divided into two beds (A and B) with a thickness ratio of 3:1. In the Carrickfergus district these units have not been adequately defined, although suggested correlations are indicated on Figure 18. The Altachuile Breccia has been recognised in the railway cutting at the northern entrance to the tunnel at Whitehead, and at Cloghfin, where the member has a thickness not exceeding 11 m. An horizon of glauconitised chalk pebbles occurs within Ballintoy B at the Ballycarry Quarry (Figure 18). Fletcher (1978, fig. 8) suggested that this horizon was related to the North Antrim Hardgrounds event of the North Antrim depositional basin, ie intra-Glenarm Chalk, but is now clear that the horizon in question lies *below* rather than above the Altachuile Breccia.

Very large pachydiscid ammonites typical of high Bed A horizons at Ballintoy Harbour can be seen on the foreshore rock-platform at Whitehead. Specimens of *Echinocorys* ex gr. *conica* collected low in Ballintoy A at Whitehead and elsewhere correspond to the Ballintoy A1 assemblages of the type-area.

GLENARM AND GARRON CHALK MEMBERS

The presence of these members is indicated by the thickness of preserved strata above the Altachuile Breccia, and by the characteristically large flints near the contact with the overlying 'Clay-with-Flints'. The boundary between the members has not been accurately located. There appears to be no intra-Glenarm non-sequence marked by erosion surfaces with concentrations of glauconitised chalk pebbles equivalent to the North Antrim Hardgrounds. An horizon of large *Echinocorys* noted in the lower part of the Glenarm Chalk in the Whitehead railway cutting presumably corresponds to the *Echinocorys* Bed with *Echinocorys* cf. *ovata* present in the Sheet 7 localities.   TPF, CJW

CHAPTER 9

Tertiary

ANTRIM LAVA GROUP

In early Tertiary times, there was extensive volcanic activity in the North Atlantic region. 'Flood basalts' poured out, inundating the adjacent land areas, probably soon after the Cretaceous rocks were raised above sea level, and plugs, dykes and sills were intruded. Subsequently, late-Tertiary and post-Tertiary denudation substantially reduced the area, and thickness, of the basalt outcrop but even so, the Antrim Lava Group still covers about 3800 km^2 of northeast Ireland.

In this area the thick Interbasaltic Formation, which occurs extensively elsewhere in County Antrim, is not present and all the lavas are thus assigned to the Lower Basalt Formation.

Within the area the lavas are up to 250 m thick and there is evidence, from zeolites (Walker, 1960a, p. 524), that, in some places, up to 450 m of lava has been removed by erosion.

In this area, the basalts always rest on an eroded surface of the Ulster White Limestone Formation but are separated from it by a thin residual soil of flint nodules and clay, the 'clay-with-flints' bed. These flints are commonly red-stained, probably as a result of subaerial, tropical weathering (Lamont, 1946), or due to baking by the lavas.

Extrusion of the lavas was intermittent and the weathering of earlier flows produced inter-lava beds of bole, sometimes containing vegetation debris including trees which were engulfed by later lava flows and are now preserved as casts (Walker, 1962a, p. 5).

The dominant basalt type is alkaline olivine-basalt, although one flow of a big-feldspar basalt has been recorded near Beltoy and a flow-banded trachyte occurs on Slate Hill.

Most of the lavas are of the massive aa type with a compact central core and a slaggy top and base. However, there are also examples of pahoehoe-type, each made up of a number of thin flows units (Plate 14).

The flows are extremely variable in thickness and are demonstrably lenticular, so that individual flows can seldom be traced for more than a few kilometres, even in the well exposed coastal sections or in the face of the escarpment.

There is evidence, in the cliffs at White Head and Black Head (Plate 15), of small scale explosive activity.

Pipe-amygdales occur in many flows, particularly at Black Head where, restricted to the lower part of the flow, they may be short and bent in the direction of flow of the lava, or may be long and straight, extending from the bottom to the top of a flow. These pipe amygdales originated where hot lava flowed over wet ground and bubbles of steam were generated, which then rose through the viscous lava.

The tops of most flows are vesicular and some are vesicular throughout. The bubble-cavities are all, like the pipe amygdales, filled with zeolites and calcite.

The age of the igneous activity is not known with precision, since only fragmentary plant material is available for palaeontological dating. Early workers regarded the Interbasaltic flora as Miocene or Eocene whilst work on pollen from Mull and Antrim (Simpson *in* Eyles, 1952, p. 4), indicates a late Miocene or early Pliocene age. Watts (1962, p. 600: 1963, pp. 117–118) gives a late Eocene or early Oligocene age to the Lough Neagh Clays which overlie the lavas. Geochronological work, using the K/Ar method, has produced somewhat contradictory results because of the difficulty in getting fresh and zeolite-free material. The consensus figure gives an absolute age of about 59 million years, ie Palaeocene.   EJC, HEW

Plate 14   Thin lava flows of the
Lower Basalt Formation. Black Head   [NI 366]

Plate 15 Pipe amygdales in the central position of a lava flow. Black Head   [NI 365]

PETROGRAPHY

The lavas are nearly all holocrystalline rocks, consisting essentially of plagioclase, augite, olivine and opaque oxide; chlorite and zeolite also occur in many of the rocks. Porphyritic basalts are common with olivine the most abundant phenocryst; plagioclase phenocrysts are less common and augite phenocrysts are quite rare. Some olivine-porphyritic basalts have no groundmass olivine but in general the mineral species of the phenocrysts occurs also in the groundmass—albeit of slightly different composition. Few rocks are completely fresh, the most common alteration products being chlorite, zeolite and clay minerals; some of these minerals also occur as late-stage primary minerals filling interstices in the rock and coating vesicles.

The most common mineral, plagioclase, ranges in composition from An_{73} to An_{25}. Complex oscillatory zoning occurs in most phenocrysts and some show a patchy zoning which suggests that periods of resorption interrupted growth of the crystal. The groundmass plagioclase crystals generally have calcic cores normally zoned to more sodic rims although some rocks contain unzoned crystals in the oligoclase-andesine range. Most of the crystals are 0.1 to 0.2 mm long with a length/breadth ratio ranging from 1.5 to 4.0 (average about 2.5). They are generally subhedral although late interstitial growth has modified the shape of many laths. This modification is lacking when laths are included in zeolite (NI 3598) or is hardly apparent when the sheer abundance of crystals forms a felt which in some rocks is unoriented (NI 2066) and in others is highly directional (NI 1264).

E. J. Cobbing noted convolute swirls of plagioclase laths around olivine phenocrysts in specimens NI 1258 and NI 1259. In very fine-grained rocks the feldspar laths commonly contain small granules of pyroxene, but fresh crystals in the coarser basalts contain only small amounts of olivine, chlorite, opaque oxide or spinel (NI 3595). In 80 per cent of the rocks examined the feldspar is altered or replaced to varying degrees by zeolite, or chlorite, or both, and in parts of some basalts the plagioclase cores are sericitised (NI 3597). The plagioclase content of the basalts ranges from 30 to 75 per cent with an average of 50 to 55 per cent, and because plagioclase is the commonest mineral it has been used as a basis for a crude graphical classification of the flows. A histogram of the frequency of occurrence of basalts with different plagioclase compositions (measured in the cores of the crystals and indicated by per cent An) is given in Figure 19. The spread of compositions and their frequencies indicate that the lavas are predominantly basalts, with subordinate hawaiites, mugearites and trachytes, also that there is a continuous spectrum of compositions and not a series of lavas that fall into two or three well defined discrete categories (as might have been suggested had there been less extensive sampling).

One specimen (NI 3086) from Slate Hill [3719 9503], provisionally called a mugearite, was identified by R. A. Old as a trachyte. It consists mainly of alkali feldspar laths (90 per cent), with some 5 per cent of plagioclase, about 5 per cent of chlorite and opaque minerals and about 1 per cent of quartz. An analysis is given in Table 3 and some analyses of mugearites and trachytes from Tertiary rocks in Skye are shown also. Compared with the flows from Skye with similar silica content, the Slate Hill rock has slightly higher ferric iron and lower soda.

About 60 per cent of the lavas examined contain olivine phenocrysts. In thin section they are seen to occur singly or in clusters and although some are euhedral most are sub- or anhedral, their shapes having been modified by adjacent olivine grains or by the surrounding lava. Crystals entirely surrounded by finer grained basaltic groundmass commonly have rounded edges and embayments and in places have rounded inclusions. However only a few crystals have reaction rims of orthopyroxene and some of these are discontinuous (NI 1254). Groundmass olivines are generally enclosed in clinopyroxene or plagioclase and appear to be without reaction rims. In some rocks olivine phenocrysts form centres around which plagioclase laths exhibit fluxion structure (NI 1258) whereas in others they are set in clear pools of chabazite (NI 1257). In one specimen shown in Plate 16c (NI 3607) olivine and zeolite form a coherent oval mass which is surrounded by plagioclase laths showing fluxion structure. Most olivine grains are altered but the degree of alteration ranges from none at all (NI 3609) to complete (NI 1264), the product being predominantly an olive brown montmorillonite or saponite. Yellow green serpentine and green chlorite are other common but less abundant products of alteration, and in a few rocks the olivines have bright orange alteration bands. In NI 1264 pseudomorphs after olivine consist of chlorite and calcite.

Table 3 Chemical analyses of lavas

	I	A	B	C
SiO_2	64.00	49.68	58.64	66.13
Al_2O_3	14.34	16.99	16.38	16.03
Fe_2O_3	5.60	3.45	3.05	3.17
FeO	1.48	8.99	4.91	0.70
MgO	0.47	2.79	1.06	0.84
CaO	2.70	5.46	2.90	1.45
Na_2O	3.77	5.78	6.07	5.34
K_2O	3.71	1.90	3.49	4.82
H_2O+	1.27	1.77	0.55	0.36
H_2O-	1.36	0.34	0.99	0.43
TiO_2	0.67	2.13	0.89	0.61
P_2O_5	0.26	0.48	0.66	0.08
MnO	0.11	0.27	0.18	0.10
CO_2	0.01	—	—	—
Ba	0.09	0.12	0.13	—
F	0.03	0.23	—	—
Zr	0.07	—	—	—
Total	99.94	100.40	99.90	99.77

I Trachyte, outcrop in track 400 m WNW of Ballyvallagh Bridge, Slate Hill, Antrim. NI 3086. (Analysts: G. A. Sergeant, J. I. Read, J. M. Murphy and D. R. Powis.)

A Mugearite from the type locality. (Analyst: J. H. Scoon *in* Muir and Tilley, 1961.)

B Mugearite-trachyte, Totardor, Skye. (Analyst: J. H. Scoon *in* Muir and Tilley, 1961.)

C Trachyte, Ros a 'Mheallain 1½ miles NE of Bracadale, Skye. (Analyst: J. H. Scoon quoted *in* 'The Geology of Northern Skye' *Mem. Geol. Surv. G.B.*, 1966.)

Lavas with olivine phenocrysts O and zeolites Z
Olivine phenocrysts with spinel inclusions Ө
Plagioclase anorthite content not measurable in four samples
Total number of samples 46

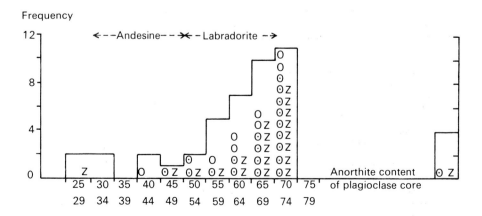

Figure 19 Histogram of basalt lavas with different plagioclase core compositions

In two-thirds of the rocks with olivine phenocrysts, the phenocrysts contain octahedra and subhedral grains of brown spinel (see Plate 16E). One specimen (NI 1258) has plagioclase with brown spinel inclusions up to 0.15 mm across adjacent to clusters of olivine grains. Normally the spinels are brown and isotropic averaging about 0.005 mm in diameter, but some of the larger grains outside the olivine phenocrysts have brown cores and opaque rims. Analyses of the translucent brown spinels included in olivine and of one spinel with an opaque rim lying in plagioclase are given in Table 4, and the results of electron probe scans across the grains carried out by Mrs A. E. Tresham on a Cambridge Instruments Geoscan II are shown in Figure 20. The analyses were recalculated into unit cell formulae, assigning Si to olivine, Ti to ulvospinel (Fe_2TiO_4) and the remaining elements to spinel. The amount of Fe_2O_3 was calculated assuming the divalent and trivalent elements to be present in the spinel in their stoichiometric proportions. The resulting cation proportions indicate that the brown translucent grains are chromian spinel (following the terminology of Stevens, 1944) and that the opaque rims to spinel grains outside olivine are a mixture of about 60 per cent chromian

spinel and 40 per cent ulvospinel. The analysis of each spinel has been positioned next to the analysis either of its host olivine or of an adjacent olivine except for analysis III (an olivine with no spinel inclusions) and analysis VIII—a groundmass olivine which probably crystallised contemporaneously with the dark rim of the spinel. The olivine analyses were placed in order of decreasing Mg/Mg+Fe ratio and this trend is followed by the spinel analyses. The olivine phenocrysts are normally zoned to varying extents, the most magnesian core analysed being $Fo_{87}Fa_{13}$ and the most iron-rich rim being $Fo_{71}Fa_{29}$ with the groundmass olivine more fayalitic at $Fo_{65}Fa_{35}$. In Figure 20 two spinel grains and the variation in their chemical content are shown from specimen NI 1258. The trace for each element represents the X-ray response at a particular wave-length and its height is a qualitative but not an accurate quantitative indication of the element concentration. Thus Mg, Cr and Al are concentrated in the spinel cores while Fe is concentrated in the opaque rims; the Fe and Al distribution also might suggest slight zoning within the spinel cores.

The distribution of both brown spinel and opaque inclusions is common but sporadic in density. Other minerals found within olivine grains may represent embayments rather than inclusions and many examples of chlorite and zeolite 'inclusions' are almost certainly of this type. However, the crystal form of some inclusions of zeolite and chlorite suggests that they are completely surrounded by olivine and one example of ?mesolite in NI 3608 could be of this type. Curiously, the olivines in NI 3609, already mentioned as being unaltered along rims and cracks, contain round inclusions of chlorite. Brown spinels and opaque grains commonly occur in chlorite pseudomorphs after olivine, apparently unaffected by the alteration (NI 1263).

Clinopyroxene is present in all except two of the rocks examined and constitutes between 5 and 50 per cent of the groundmass minerals. It commonly occurs as ophitic or subophitic pale brown grains (see Plate 16A) including small olivine grains as well as plagioclase and iron oxide (NI 3606); chlorite or zeolite are rare inclusions. In about 20 per cent of the lavas the habit of the pyroxene may be described as granular with subophitic tendencies; in some places the grains have shadowy extinction under crossed nicols and this may indicate slight compositional zoning (NI 3595). The pyroxene grains in NI 2062, NI 3595 and NI 1263 are pale pinkish lilac in thin section, and in NI 451 and NI 2073 they are a strong lilac colour. In over two-thirds of the rocks the pyroxene is very fresh, even in some rocks with extensively altered olivine and plagioclase. In the remaining rocks alteration to chlorite has taken place but commonly its development is patchy and closely related to zones of chloritisation or to veins of chlorite, zeolite or calcite.

Magnetite, ilmenite and sparse sulphide make up the rest of the opaque minerals. Typically they are distributed throughout the lavas in clusters of subhedral grains, although some show good crystal shape and some are anhedral and obviously interstitial. They comprise on average about 3 per cent of the rock but Mr J. W. Arthurs recorded more than 20 per cent of opaque minerals in some samples (NI 3596, NI 3601); sampling is not sufficient to indicate whether the whole lava had these contents of iron

Table 4 Electron probe analyses of olivines and spinels in basalts

	I	II	III	IV	V	VI	VII	VIII	IX
SiO_2	39.6	0.2	39.1	38.5	0.3	38.6	—	37.1	0.7
Al_2O_3	—	35.8	—	—	33.9	—	32.5	—	5.3
Fe_2O_3	—	5.0	—	—	7.5	—	7.3	—	15.7
FeO	14.4*	13.6	15.7*	21.1*	18.5	22.0*	19.8	29.7*	36.8
MgO	44.6	15.3	43.7	39.7	12.4	38.8	11.2	32.1	5.2
CaO	0.3	—	0.3	0.4	—	0.4	—	0.4	0.2
TiO_2	—	0.8	0.1	—	0.7	0.1	0.7	—	14.6
Cr_2O_3	—	26.9	0.1	0.1	26.8	0.1	27.6	—	18.3
MnO	0.2	0.3	0.3	0.3	0.3	0.4	0.4	0.4	0.5
NiO	0.3	0.2	0.2	0.2	0.2	0.2	0.2	0.1	0.2
CoO	—	0.1	—	0.1	0.1	—	0.1	—	0.2
ZnO	—	—	—	—	0.2	—	0.2	—	0.2
V_2O_3	—	0.2	—	—	0.1	—	0.2	—	—
Total	99.3	98.3	99.4	100.4	101.0	100.6	100.1	99.9	97.8
Fo content	85	—	83	77	—	76	—	66	—
Mg/Mg+Fe	—	0.67	—	—	0.55	—	0.50	—	0.20

*Total iron calculated as FeO; Fe_2O_3 content of spinel calculated assuming stoichiometry of the grain. (Analyst: Mrs A. E. Tresham.)

I   Olivine 'B' from basalt NI 1258, 1 km NW of Beltoy on Glenoe Road, Antrim [J409 954].
II   Spinel included in olivine 'B' in NI 1258 (see I).
III   Olivine from basalt NI 3598, from outcrop on road 200 m at 100° from Moat Farm, Antrim [J402 953]. No spinels were found in this rock.
IV   Olivine 'A' from basalt NI 1258 (see I).
V   Spinel included in olivine 'A' (see IV and I).
VI   Olivine 'C' from basalt NI 1258 (see I).
VII   Core of zoned spinel adjacent to olivine 'C' (see VI and I).
VIII   Olivine 'G' from groundmass of basalt NI 1258 (see I).
IX   Rim of zoned spinel adjacent to olivine 'C' (see VI and I).

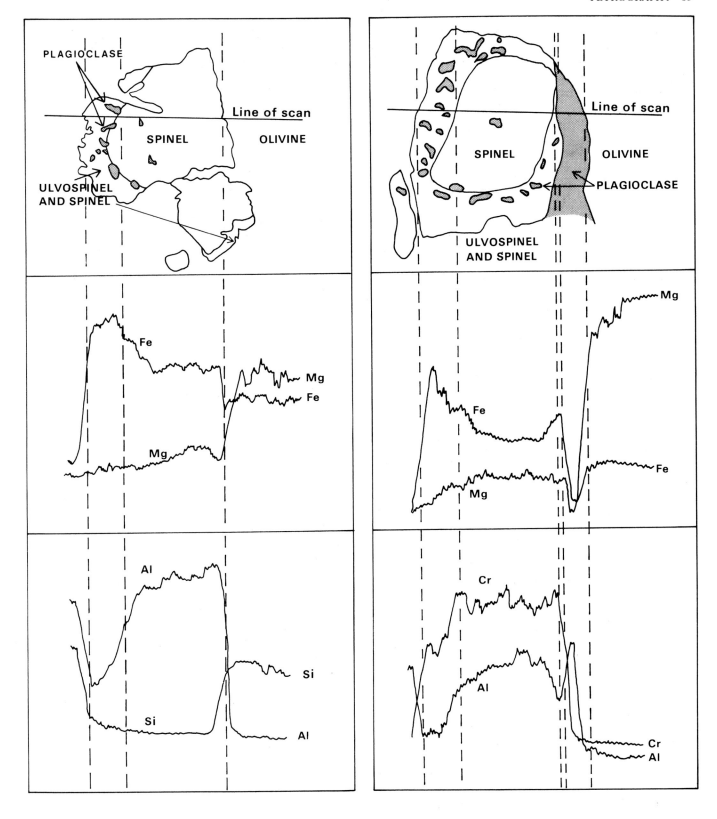

Figure 20 Variation of some elements in spinel and olivine as determined by electron probe

oxide or whether they are local segregations. In the rocks with low contents of opaque minerals the main species is magnetite (or hematite) and it is associated with alteration of olivine and pyroxene; its form is generally anhedral and in places skeletal (NI 3597). In one specimen (NI 1259) the oxides included in chabazite are euhedral. Many of the magnetite grains are altered at their edges to hematite (NI 1257, NI 3596) and in places adjacent chlorite is reddened.

Chlorite is present in all but two of the lavas examined. Most occurs interstitially in the groundmass of the rocks and in amygdales; the rest is an alteration product of olivine or pyroxene, or occurs in cracks in plagioclase crystals. In general the chlorite is pale to medium green in thin section, slightly pleochroic, with low, locally 'anomalous', birefringence and is either fibrous in habit or extremely fine-grained. In places next to iron oxides or in concentric rings in amygdales, the colour of the chlorite can vary between, orange, brown, green, yellow-green and very pale green (NI 3596). In NI 3606 the cores of the plagioclase crystals have been altered to chlorite whereas the rims remain relatively clear, and in NI 3601 the cracks in some olivine grains contain chlorite with well developed radial growth structures. Another alteration product of olivine consists of green, olive or brown bladed crystals of montmorillonite or saponite. Their orientation in the olivine grains is inconsistent but commonly their long axes are perpendicular to cracks in the host grain and they project into the body of the grain like canine teeth. The montmorillonite has a lower mean refractive index than chlorite and a higher birefringence but some golden brown material has a lighter mean refractive index than chlorite and this shows a somewhat silky aspect in the extinction position (NI 1257, NI 1258 and NI 1260). It is not as common as chlorite, occurring in just over half the rocks examined. The fourth alteration product of olivine, serpentine, occurs in only a quarter of the slides, typically as fibrous green or yellow-green crystals (NI 1263). In contrast to the usual order of alteration of the primary minerals of the lavas it may be noted that in NI 1256 the clinopyroxene has been converted to chlorite whereas the olivine remains fresh.

Optical microscope techniques are not good enough for rigorous identification of zeolites—this is best done by X-ray diffraction methods—and the following observations are of a general nature. The occurrence of zeolites in the lavas has been plotted in relation to the plagioclase composition of each lava (Figure 19) and from this it is apparent that the zeolites occur over the whole range of plagioclase compositions with no preference for any particular value, although sampling of lavas with sodic plagioclase is too limited to draw firm conclusions. Chabazite is the most common zeolite, occupying interstitial positions in the fabrics of more than one-third of the lavas examined (Plate 16B). In some of the rocks it is associated with porphyritic olivine and in places contains small crystals of euhedral plagioclase and olivine, and anhedral augite (NI 3607 in Plate 16C, and NI 3609). This indicates that the chabazite in the groundmass probably formed in the later stages of crystallisation of the rock and not after the lava had solidified. Natrolite, thomsonite, stilbite, mesolite and ?heulandite are all less common than chabazite, various

combinations occurring in about 15 per cent of the lavas. It is rare to find two species of zeolite in contact in one thin section and in NI 1261 which has about 8 per cent chabazite and 5 per cent ?mesolite only one contact between the two was found. Development of zeolites is apparently discrete on a small scale and it may indicate that multiple small zones, less than 5 mm in diameter, and of variable chemical character were established late in the period of solidification of the lava. Zeolites tend to be scarce in the vicinity of chlorite-filled amygdales but are commonly adjacent to secondary chlorite derived from olivine or unknown groundmass material. Chabazite contacts with plagioclase are well defined (NI 1257) and it does not seem to replace plagioclase, but much feldspar is cracked and net-veined and converted to varying degrees by one or two other zeolites—probably thomsonite or natrolite (NI 1257, NI 1260, NI 1263, NI 3609).

Apatite, of stumpy prismatic habit, was found in only two lavas. In one (NI 1254), the grains are euhedral or subhedral in shape and fresh, but in the other (NI 1260), although the prisms looked fresh, sections across the c-axes showed many to have hexagonally shaped cores of pale green chlorite. Calcite is of sporadic occurrence in the lavas both as a replacement mineral of olivine and groundmass material and as thin veins.

RRH

Plate 16 Photomicrographs of basalt lavas

A. Basalt (NI 3606), showing ophitic clinopyroxene in some parts of the groundmass (centre and right) with chlorite and opaque oxides interstitial elsewhere [J 380 944] × 2.5

B. Basalt (NI 1261). Olivine-basalt in north-east and south-west corners surrounds patch of zeolites which consist of chabazite (middle top) in contact with ?mesolite (fibrous, lower right) [J 404 957] × 8

C. Basalt (NI 3607), showing broken olivine phenocrysts in a matrix of chabazite, feldspar laths and small grains of olivine. Between this assemblage and normal basalt is a mantle of roughly aligned feldspar laths [J 387 957] × 3

D. Basalt (NI 3601). The large crystal is a broken olivine phenocryst which has been altered to chlorite and hematite along its rims and cracks. Radial and concentric growth of olivine is present in the lower olivine fragment, and in the lower left of the picture plagioclase cores have been replaced by chlorite [J 396 914] × 2.5

E. Basalt (NI 1258), showing a group of spinel crystals, the largest of which is 0.15 mm across. This has a core of chromian spinel and a rim of chromian and ulvospinel. It is set in feldspar next to an olivine phenocryst [J 409 954] × 1.5

F. Basalt (NI 1258). In the centre of the picture is a zoned spinel with the translucent part (chromian spinel) embedded in the olivine phenocryst and the opaque part (made opaque by a significant content of Ti) next to the basaltic groundmass [J 409 954] × 1.5

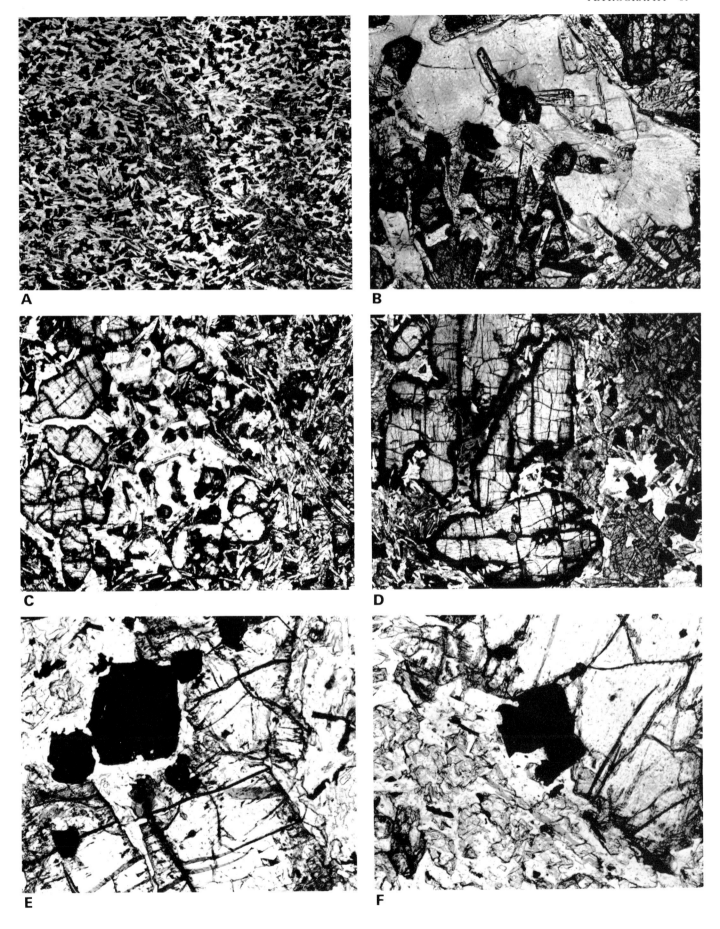

CHAPTER 10

Intrusive igneous rocks

Intrusive rocks of two generations are known in this district. Lamprophyre dykes associated with the Caledonian Newry Granodiorite occur in the Lower Palaeozoic rocks of County Down, while Tertiary dolerite intrusions, mainly in the form of dykes, are most common in the Mesozoic rocks on the north side of the Lough but are also present in the area to the south.

CALEDONIAN INTRUSIVE ROCKS

The lamprophyre intrusions are restricted to the Lower Palaeozoic outcrop to the south of Belfast Lough, where they occur both as true dykes cutting across the bedding planes of the sedimentary rocks and as sills intruded conformably, or nearly so, with the bedding.

Where the dykes cut across the bedding they are easily recognised but recognition of the sills can be difficult as they resemble the massive greywacke units in colour and blocky aspect.

No lamprophyres have been recognised in inland exposure and the 15 which are recorded all occur on the wave-cut platform around the coast.

During the original geological survey a few of the lamprophyre intrusions were recorded and referred to in the Memoir (Hull and others, 1871, p. 14) as felstones. Later McHenry and Watts (1898, pp. 74–75) classified these as lamprophyres of the minette and kersantite groups. A year later Seymour (1899, pp. 180–181) reclassified the lamprophyres as kersantites and camptonites, because he considered the dominant feldspar to be plagioclase and the dominant ferromagnesian mineral to be sometimes biotite, sometimes hornblende and rarely augite.

Reynolds (1931, pp. 98–99) divided the lamprophyre dykes, lying to the south of this sheet, into two series; the older consisting of crushed minettes and, rarely, vogesites and a later series of uncleaved lamprophyres in which hornblende, or biotite, or monoclinic pyroxene is predominant. The lack of cleavage in the younger series was attributed to the dykes having been intruded after the imprint of the regional cleavage.

Although the Ordovician sediments are more intensely cleaved than the Silurian strata the dykes in the Ordovician are not noticeably more cleaved than those in the younger rocks, except for two dykes (NI 1344) lying some 580 m N of Orlock Bridge where cleavage is well developed in the dyke rock and the adjacent sediments.

The most spectacular dyke occurs on the foreshore due east of the mouth of Galloways Burn where it forms a 0.6-m thick wall standing up to 0.5 m above a flat-lying, wave-smoothed bedding plane in silty mudstones.

The dykes are best developed on Copeland Island, where one multiple intrusion can be traced for 460 m along the south shore of Barnagh Bay and is off-set by numerous small faults (Figure 21). Unlike the majority of the sills in the area, which tend to be fairly regular, an intrusion at Collins Port also on Copeland Island is remarkably sinuous in plan with numerous small off-shoots, reflecting the less competent nature of the predominantly argillaceous beds into which it was intruded.

Petrology

The lamprophyre dykes are all spessartites but may be divided into three distinct textural groups. They are almost certainly all derived from a single magma-type. None, however, contain clear evidence relating to the origin of this magma.

Fifteen of the rocks (NI 1339–43, 1346–51, 1355–6, 1358–9) are typical, medium grey-coloured, spessartite lamprophyres. Grain-size variations are dependent on the thickness of the intrusive bodies or on the positions of samples with respect to the margins of dykes and sills. Some rocks have a marked fluxion texture (NI 1346, 1351); others show a more random disposition of constituent minerals suggesting less rapid flow consolidation (NI 1343, 1347). Alteration of the primary minerals is common, although variable in intensity. This alteration was probably due, partly to autometasomatic processes, but may also reflect changes induced by the low-grade regional metamorphism.

The spessartites contain phenocrysts of hornblende (partly or wholly altered to various mixtures of carbonate, chlorite, iron-oxides, clay minerals and quartz), biotite (typically pseudomorphed by chlorite, carbonate and iron-oxides, or mixtures of these minerals) and sporadically plagioclase (oligoclase) set in a matrix dominated by either tabular (NI 1339, 1341–3) or lath-shaped (NI 1340, 1356) crystals of oligoclase. Sodium cobaltinitrite staining in specimen NI 1347A indicates that small amounts (3 to 5 per cent) of K-feldspar are developed at the margins and, locally, in the interiors of oligoclase crystals in the matrix. The K-feldspar occurs as scattered granules, or in dense clusters of granules seemingly replacing the plagioclase. Unfortunately, the mineral cannot be detected without the aid of staining, so even though the staining of a second, plagioclase-rich rock (NI 1345A) yielded a negative result, K-feldspar may be present in some of the other lamprophyre rocks.

In addition to feldspar(s), the matrices of the spessartites contain scattered biotite (commonly altered to chlorite and/ or carbonate), ovoidal carbonate/chlorite/iron oxide-aggregates replacing amphibole, quartz anhedra, some interstitial chlorite, illitic and kaolinitic clay minerals derived mainly from feldspars, abundant opaque oxide and accessory apatite. The oxide in certain specimens (NI 1357, 1359) is white in reflected light suggesting leucoxenised ilmenite, but in most it appears to consist of dark red or brown material which is probably mainly goethite with some hematite.

The secondary carbonate in seven of the rocks (NI 1339–43, 1351, 1358) is dolomite and most of these hand specimens are a distinctly lighter buff colour. Some also show reddening due to dispersed iron oxides. The carbonate in the remainder of the samples is mainly calcite.

A few rocks are cut by carbonate and/or quartz veinlets (eg NI 1343, 1351, 1358–9). Two specimens contain sedimentary xenoliths. The first example (NI 1342), bears a small xenolith of fused sandstone which is at least 1 mm × 3 mm × 3 mm in size. Abundant dolomite occurs in the lamprophyre adjacent to this inclusion. In the second specimen (NI 1356), the inclusion (at least 2 mm × 2.5 mm × 5 mm) is a carbonated silty sandstone into which the lamprophyre fingers. These xenoliths suggest that the quartz anhedra in the matrix could be due in part to the assimilation of sedimentary material. Some of the quartz, notably that associated with altered amphibole phenocrysts (NI 1358), clearly results from the breakdown of the primary mineral constituents.

Because of the erratic distribution of phenocrysts and the secondary alteration, no modal analyses have been performed. However, visual estimates (ignoring secondary minerals) indicate the following approximate mineralogical composition for the spessartites:

	Volume per cent
oligoclase (+K-feldspar where present)	55–75
hornblende	10–25
biotite	10–15
quartz	1–5
iron oxides	3–5

In the highly altered rocks, no biotite or amphibole remains; chlorite and carbonate may form between 20 and 45 per cent of such material.

The grain sizes of the mineral constituents vary considerably. In the finest grained rocks, which are either from thin dykes and sills or from the margins of thicker intrusions, the average grain size of matrix constituents may be below 0.05 mm (eg NI 1339–41, 1346, 1351). In the coarser rocks, the average grain size is around 0.1 mm (eg NI 1343,

1347–9). Commonly phenocrysts are below 1.0 mm in size. The largest amphiboles are up to 3.0 mm long, biotite flakes reach 5.0 mm and plagioclase crystals may be 4.0 mm across.

Four samples in the present batch do not fit in precisely with the above generalised description of the lamprophyres. Three of these (NI 1344, 1352, 1357) contain abundant, fairly fresh flakes of biotite (20 to 30 per cent by volume) and resemble kersantites. NI 1352 bears a 3 mm-long quartz-carbonate clot (probably a sandstone inclusion), around which small, fluxion-orientated biotite crystals are clustered. Apart from the enrichment in biotite, these rocks resemble those of the main group. Secondary carbonate in specimens NI 1344 and 1352 is calcite; in NI 1357 it is dolomite and the hand specimen is buff-grey in colour.

The one remaining sample (NI 1345, from the centre of a sill) is an oligoclase-rich rock bearing scattered quartz anhedra, interstitial chlorite and minor carbonate (dolomite), with abundant grains and aggregates of limonite and iron oxide (probably mainly goethite) and could be described as an oligoclasite. Sodium cobaltinitrite staining indicates that no K-feldspar is developed at the margins of feldspar crystals. No recognisable pseudomorphs after biotite or hornblende occur in the rock, which presumably comes from a feldspar-rich layer or segregation within a fairly typical spessartitic intrusion. AEG, JRH

TERTIARY INTRUSIVE ROCKS

The eruption of the Antrim lavas was accompanied and succeeded by a variety of intrusive phenomena, particularly volcanic plugs and dykes which may, in some cases, have acted as feeders to the lavas. The only plug in the area is at Carneal [389 959] and, in being intruded through the lavas this plug has dragged up a screen of highly metamorphosed chalk. Details of the Carneal Plug and its remarkable mineralogy are given in Chapter 11.

Basic dykes are very common, over one hundred being recorded, mainly in foreshore exposures. As in the Belfast area to the south-west, the dykes are most abundant in the Triassic rocks, particularly in those of the Mercia Mudstone Group, where they are often very irregular in direc-

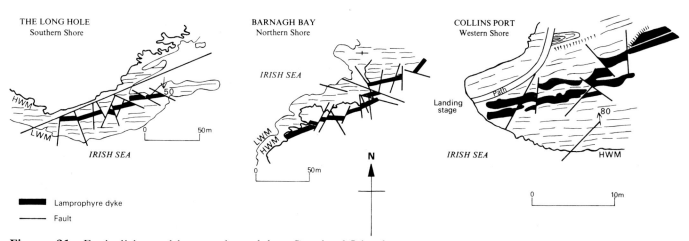

Figure 21 Fault-dislocated lamprophyre dykes. Copeland Island

tion and in thickness. They appear to be less common in the Ulster White Limestone and in the basalt lavas, where they are less easily discerned, and are certainly much less common in the Palaeozoic rocks where they are usually massive and regular in direction.

The most notable exposures of dykes are on the foreshore between Greencastle and Carrickfergus where a complex swarm of dyke-like intrusions stands out from the sea-washed mudstones. The dykes are frequently sinuous and bifurcate, have cross-connections with one another and occur as a series of *en-échelon* bodies, though possibly connected at depth. Sill-like cross connections are also developed and these are well exposed at Greenisland (Figure 22).

Intrusion breccias are developed both in the igneous intrusions and in the Triassic mudstone wall-rock. They are most common where the intrusions are complex and especially so where sills are present. The dolerite breccias comprise angular fragments of dolerite set in a matrix, often slickensided, which usually separates the blocks and is rich in clay minerals and carbonate. Though normally only marginal these dolerite breccias can occupy the full thickness of the intrusion.

The wall-rock breccias are more common than the dolerite type but are usually only a few centimetres thick. They consist of red and grey mudstone fragments and may be in sharp or gradational contact with the dolerite breccias.

The breccias were apparently formed after the consolidation of the intrusions. The matrix of the dolerite breccias is not solidified magma, as it would be had the brecciation taken place before complete consolidation, and in the

Greenisland Sill, below Raven Hill, the strong jointing in the main sill is matched by jointing in disorientated blocks in the breccia. The brecciation was probably caused by a build-up, and violent release, of superheated volatiles which would have become trapped, especially under the sills.

The massive intrusion [415 873] on which Carrickfergus Castle is built appears to be a local thickening of a dyke, perhaps associated with a fault, similar to the Tertiary intrusion in the Lower Palaeozoic rocks at Ballymoney [434 787], north of Craigantlet. Another exceptionally massive dyke forms The Briggs [467 901] west of Kilroot.

The general trend of nearly all the Tertiary dykes in Antrim, is north-north-west and in general they are less than 3 m thick.

Charlesworth and Hartley (1935, p. 193) state that the Carrickfergus–Lisburn dyke group is part of the Tardree 'swarm'. Of the large number of dykes on the north side of the Lough, Walker (1959a) has pointed out that the crustal expansion in that area is of the order of 4 per cent. Walker's suggestion (1959a, p. 195) of an earlier phase of intrusion prior to the main swarm has not been proved.

It has been suggested that some of the dykes may have been fissure-eruption feeders and Patterson (1950) and Walker (1959a) have described examples from Browns Bay to Skernaghan Point on Island Magee. Walker has expressed dissatisfaction with the evidence and Thompson (*in* Preston, 1971, p. 90) has concluded that these and other examples in the same region are fissures infilled from above.

The only evidence of major explosive volcanic activity in the area is a volcanic agglomerate penetrated by a borehole near Kilroot Church [452 896]. The blocks in the vent

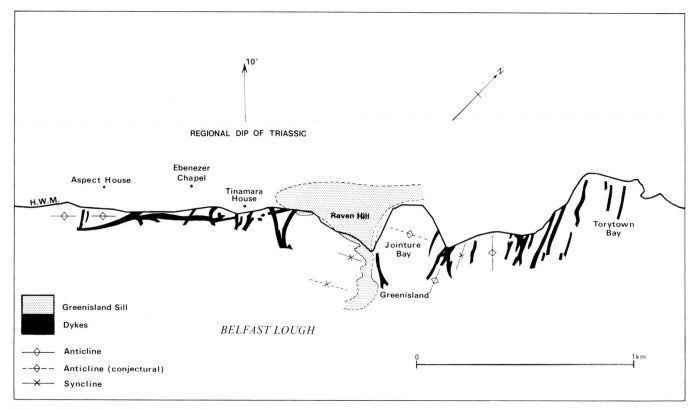

Figure 22 Plan of the complex intrusions on the foreshore at Greenisland

include a limestone, probably of Carboniferous age, and the indications are that this is an explosion vent similar to many known elsewhere in Antrim. A similar agglomerate, of basalt blocks, is seen over an area of a few square metres on the foreshore at Cloghfin Port (Figure 12).   HEW, EJC

Petrology

The basalts and dolerites are all petrographically similar, a few being fresh and unweathered but most showing partial or complete serpentinisation of olivine, and variable chloritic alteration of the groundmass. Many are porphyritic with plagioclase, olivine or clinopyroxene occurring in two generations, but this is not usually obvious in hand specimens, which show only dark grey or black rocks, rarely vesicular and commonly fine-grained.

In thin section most of the rocks examined proved, using the criterion of 0.5 mm maximum length of plagioclase laths in the groundmass, to be basalts, but some are true dolerites, a particularly fresh example being the rock (NI 1278) below Carrickfergus Castle [415 873].

Plagioclase is commonly zoned from a basic labradorite core to an andesine rim but some of the larger porphyritic crystals show complex oscillatory zoning. The phenocrysts reach 3.5 mm in length (NI 1272), and glomeroporphyritic clusters of feldspar range up to 4 mm across (NI 1277, 1282). Both albite and carlsbad twins are commonly developed and in some rocks there is a good development of cruciform twins. The length to breadth ratio of the plagioclase crystals ranges from one to more than twenty, the laths with high ratios commonly being bent or transversally cracked (NI 1273). In most rocks the plagioclase is very fresh but in some variable alteration and replacement by chlorite, clay mineral or zeolite occurs.

Olivine occurs in many of the rocks as small subhedral grains, serpentinised along cracks and margins of the grains, and less commonly, as large, variably altered phenocrysts up to 3 mm long. In specimen NI 1274 from Thompson's Point [393 863] olivine occurs in glomeroporphyritic aggregates with clinopyroxene and in specimen (NI 1282) from Raven Hill [383 853] it has been protected from alteration to serpentine, in contrast to the groundmass olivines, by the armouring effect of the surrounding plagioclase phenocrysts. With chlorite and serpentine common in the rocks it is difficult to determine olivine percentages but amounts probably vary from zero (NI 1277) to about 30 per cent (NI 1272). Red-brown spinels are common inclusions in the olivines but their abundance is variable and some rocks lack spinel altogether: in one of these (NI 1278) small plagioclase laths are included in the olivine. Groundmass olivines are commonly extensively altered to serpentine and chlorite, and in two specimens (NI 3777 and 3778) the zonal alteration has produced layers of serpentine resembling the cross section of an onion.

In a specimen (NI 1276) from west of Thompson's Point [398 866], red-brown spinels occur as discrete crystals up to 0.3 mm across in the groundmass of the rock as well as inclusions in olivine and altered olivine phenocrysts. Typically they are 'protected' from the groundmass by a rim of

opaque oxide whereas those surrounded by chlorite and serpentine in altered olivine grains show only sporadic and incomplete development of this rim. One dolerite (NI 3776) from Cloghan Point [467 903], south-west of Whitehead, contains olivines with pale brown spinel inclusions and in a basaltic dolerite from near Thompson's Point red-brown spinels are included in glomeroporphyritic aggregates of clinopyroxene; neither pale brown spinels nor the inclusion of red-brown spinels in pyroxene were seen in the basalts of the area. The normal alteration to serpentine appears not to affect the spinels but conditions must have been slightly different in dolerites at Cloghan Point where alteration of the spinels to a pale brown chloritic mineral has occurred.

Ophitic or subophitic crystals of clinopyroxene up to 3 mm across are present in all rocks except those with intergranular and variolitic textures (NI 1282 and 3610). Glomeroporphyritic aggregates of pyroxene were observed in one rock only (from near Thompson's Point). In general the pyroxene is pale brown in thin section but some occurrences are colourless and others have quite a strong lilac tint. Sharp boundaries between fresh pyroxene and chlorite and serpentine of the groundmass are noticeable and common, but where linear (or planar) zones of alteration traverse the rocks the pyroxene is affected and breaks down to a mixture of chlorite and opaque ore granules.

Opaque oxides are ubiquitous throughout the rocks, occurring as subhedral or interstitial grains and most commonly associated with pyroxene or chlorite. In dolerites such as NI 1704 from Greenisland [387 853], two distinct forms are apparent—the granular or equant and the acicular, probably representing magnetite and ilmenite respectively. Some of the roughly equant grains in this rock enclose small laths of plagioclase, but in most other rocks the opaque minerals are free of inclusions.

Interstitial minerals in the dolerites and basaltic dolerites consist of chlorite, serpentine, calcite, zeolites, clay minerals and glass with or without crystallites. The chlorite is pleochroic from yellow-green to green to pale green and has variable birefringence although the chlorite in the amygdales tends to have lower birefringence than that occupying the spaces between plagioclase and pyroxene crystals. Calcite occurs as a general secondary groundmass mineral throughout the rock or as amygdales, commonly rimmed by chlorite. In a few rocks (NI 1275, 1699) it replaces olivine phenocrysts. Zeolites do not seem to be as common in the intrusive rocks as in the lavas, but heulandite and thomsonite have been tentatively identified in a dolerite (NI 3780) from Helen's Bay [459 832] and analcime in a dolerite (NI 1713) near the Creighton's Green Reservoir [433 787]. Glass, partly devitrified, is best shown by a dolerite (NI 3778) from Cultra foreshore [407 805] where it occurs in the groundmass and in amygdales with sinuous crystallites of chlorite. A breccia from a depth of 20 m in a borehole at Kilroot Power Station (NI 3612) consists of angular fragments of several types of dolerite both coarse and fine-grained, a few fragments of mudstone, and comminuted phenocrysts and dolerite set in a matrix of chlorite, brown, fibrous clay-minerals and carbonate.   RBH

CHAPTER 11

The Carneal Plug

The Carneal Plug [389 959], one of some thirty Tertiary dolerite plugs known to be associated with the Antrim Basalts (Walker, 1959a, p. 198; Charlesworth and others, 1960, p. 443) was found during the re-mapping of this sheet. (*Summ. Prog. Geol. Surv. G.B. for 1959*, p. 45 and Sabine, 1968). It is situated on the east side of the Carneal Water near its confluence with the Raloo Water, 1.2 km NNW of Carneal Bridge [395 947], and forms a small flat-topped knoll beside the flood plain (Plate 17).

The plug is roughly circular in plan with a diameter of 150 m and lies immediately to the west of the NNW-trending Carneal Fault, a normal fault with a downthrow to the east. There may be a parallel fault on the western side of the intrusion but the plug does not appear to be connected with this faulting. A linear hollow across the centre of the knoll may also mark another subparallel fault, later than the intrusion.

Marginal shearing was observed on the eastern side of the plug where a large xenolith of Ulster White Limestone is involved in the movement. The shearing indicates post-intrusive movement and suggests that the Carneal Plug was emplaced prior to the NW-trending faulting.

The plug (Figure 23) consists of a central core of coarse-to fine-grained dolerite with an outer screen of pyroxenitic composition which contains large inclusions of hornfelsed Cretaceous Ulster White Limestone up to 10 m across. The Ulster White Limestone is seen at outcrop near Carneal Bridge, 1 km S of the plug, but the depth at which it underlies the area of the intrusion is not known. It is likely to be of the order of a few tens of metres.

The rock of the central part of the intrusion may be generally described as a dolerite, though some specimens are of basaltic grain-size and fine-grained material, presumably chilled, is seen near the margin at the north-west

Plate 17 Carneal Plug. The low tree-covered knoll rises above the alluvial flat of the Carneal Water [NI 347]

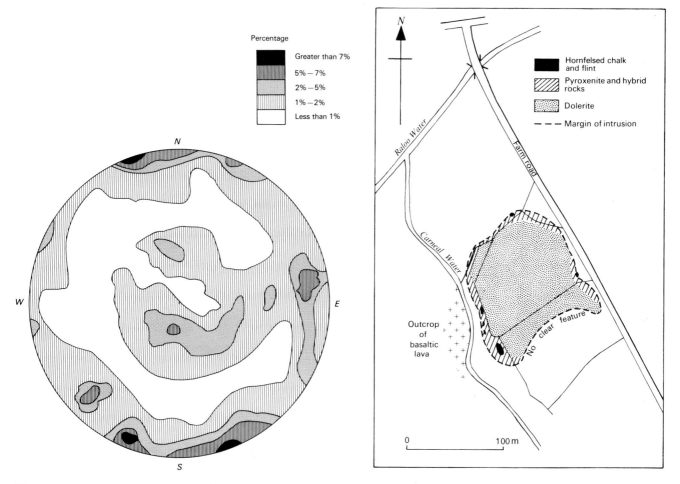

Figure 23 Plan of the Carneal Plug and stereographic plot of joint planes

corner of the intrusion, where it occurs as fragments in the dolerite. It is an ophitic olivine-bearing rock of the type common in the Tertiary intrusions of County Antrim. Exposure over the outcrop is reasonably good and the rock may be examined in small quarries and crags on the top of the knoll.

The outer screen of hybrid rocks, formed by contamination of the olivine-dolerite by limestone xenoliths dragged up by the rising magma, is seen in near-continuous, though overgrown, exposures on the north and west of the plug and in a quarry and rock knoll near the south-east corner. The zone of hybrid rocks varies in thickness from 3 to 6 m in exposed sections but may exceed this, as the contact with the basalt lavas is masked by drift.

The suite of hybrid rocks varies from slightly metasomatised basalt and dolerite to coarsely crystalline pyroxenite with individual blades of pyroxene up to 3 cm long. The coarsest grained material often occurs at the margin of limestone inclusions. Some of the fine-grained material, which looks like basalt, is seen in thin section to be a metasomatised hybrid rock. On the west face of the intrusion, about 30 m S of the point where the stream passes under the boundary wall, a small spur shows 5 m of pyroxenite with segregative clots, weathering as dark rugged knots, which consist largely of pyroxene.

The inclusions of Ulster White Limestone (Plate 18), whose reaction with the dolerite has caused the hybridisation, are best seen on the western side of the plug, where small excavations indicate attempts, at some time, to quarry the chalk, and in the area of a small quarry at the south-east corner. Most of the inclusions are small and structure has been obscured by metasomatism but in the few cases, where dips can be measured, they are vertical and near parallel to the plug sides.

The limestone has been subjected to intense thermal metamorphism and metasomatism. The larger pieces are recognisable as originally Cretaceous chalk and flint, but mineralogically the original calcite and silica have been completely replaced by a complex of calc-silicate minerals. The rock has yellow, blue and grey tints and recognisable flint nodules commonly have dark reaction rims of pyroxene and magnetite. Smaller pieces of limestone or flint can be recognised only by slight colour and texture changes on the weathered surfaces of the pyroxenite.

The high degree of metamorphism and metasomatism of the Ulster White Limestone in this occurrence must be due to the magma having been at a very high temperature during the period of intrusion. It seems improbable that the passage of molten rock continued for a very prolonged period or the vent would have been swept clear of White

Limestone inclusions, but high temperatures would doubt-less have produced the effects in a relatively short period. It is of interest to compare this vent with others in Antrim where effects on the White Limestone can be seen. Intense alteration is known at Scawt Hill (Tilley and Harwood, 1931) and Ballycraigy (McConnell, 1955), the latter near a relatively small intrusion. On the other hand the large plugs at Ballygalley Head and Bendoo produce virtually no alteration in the adjacent limestone, and must have been at low temperatures on intrusion, or have been active for only very brief periods. Alternatively, these features may have been infillings of pit craters, as seen in Hawaii, and are not conduits.

The dolerite in the core of the intrusion is well jointed and over 100 joint readings, sufficient to establish signi-ficant patterns, were taken. The contoured stereogram of the results (Figure 23) indicates the prevalence of near vertical columnar jointing and low angle joints associated with cup hollows.

The two main sets of joints are ascribed to cooling, the vertical set having developed perpendicular to the cooling surface and the shallow 'synclinal' joints parallel to the cooling surface.

Walker (1959a, p. 200) notes that some of the plugs in County Antrim have a radial-columnar jointing while others have a rather platy vertical jointing more or less parallel to the elongation.   EJC, HEW

PETROLOGY

Introduction

The igneous plug of Carneal is composed of olivine-dolerite and its variants which, as a result of contamination by chalk and flint, contain pyroxenite, hybrid melilite- and wollasto-nite-bearing rocks, and rare calc-silicates. In many of its characteristics the assemblage resembles the rocks exposed at Scawt Hill, County Antrim, which have been described in a classic series of papers by Tilley (especially Tilley and Harwood, 1931) and from which a number of new minerals were described. The mineral assemblages of Carneal, however, appear in most respects to be more restricted than those of Scawt Hill and carbonate-silicates are less well developed. Some of the more detailed aspects of the mineral-ogy have been described (Sabine, 1968, 1975; Sabine and others, 1982) and include the discovery of a new mineral, bicchulite (Henmi and others, 1973); a full account of the mineralogy and petrogenesis is given by Sabine and Young (1975). From these investigations it seems likely that the rocks were metamorphosed under the very rare conditions of very high temperature (about 1050–1100°C) and low press-ure (about 200 bars). Two suites are present: the exomorphic rocks, formed by the metamorphism and metasomatism of limestone (chalk) and flint, include rare mineral assemblages with larnite, spinel, merwinite, spurrite and scawtite; and

Plate 18   Reaction rim around a flint nodule in metamorphosed Ulster White Limestone, Carneal   [NI 349]

wollastonite, quartz, plagioclase, hydrogrossular and xonotlite are derived from flint. The endomorphic suite is of pyroxene-rich dolerite and gabbro, pyroxenite and titanaugite-melilite-rocks, and aegirine- and nepheline-bearing residua, formed by the contamination and corresponding desilication of the olivine-basalt magma. Metasomatic effects are shown in olivine-dolerite by the presence of metasomatic fronts around merwinite veins, with the production of hydrogrossular and melilite, a mechanism involving the addition of water and lime only. The replacement of fossil and mineral textures under high temperature but tranquil conditions includes the silicification of a cyclostome bryozoan, originally in calcite in flint, now preserved in wollastonite. This may represent the highest temperature to which any fossil has ever been subjected by contact metamorphism yet still retained its characteristic features.

Olivine-dolerite

The olivine-dolerite is a normal variety. It is like many of the 'plateau-magma' types of the Hebridean province and corresponds chemically fairly well with this type (Table 5). It is a dark grey rock, generally non-porphyritic, composed of olivine, pyroxene, feldspar and iron ores with minor amounts of apatite and alteration products. The analysis and norm are shown in Table 5, together with analyses of rocks from other Irish plugs; the similarity is marked. The mode of the Carneal rock shows many similarities to the norm.

Although here called 'dolerite', the rocks reveal a range of grain-size well down into the basalt group, using the criterion of 0.5 mm maximum length of feldspar lath (Wells, 1936). The coarsest type of feldspar laths commonly range up to about 0.5 mm and rarely 1.5 mm in length, with augite reaching about 3 mm across and olivine 1 mm. In the finer-grained types, feldspar laths may be only about 0.1 mm long. Olivine is chrysolite ($\alpha = 1.672$), generally in crystals that reach 0.75 mm across, rarely ophitically enclosing feldspar; some individuals are in glomeroporphyritic aggregates. The amount of olivine varies considerably, as reflected by the norms of the analysed Carneal rock (6.28 per cent) and rocks from other Antrim plugs (11.9 to 17.66 per cent; Table 5). Pyroxene, an augite of pale brown to purple but some almost colourless, may ophitically enclose feldspar. It is up to about 3 mm across, and some shows lustre-mottling on the hand specimen (NI 1602–4). It commonly accounts for about a quarter of the rock by weight; rarely there is a core of olivine. Feldspar makes up about half the rock by weight: it is mainly labradorite (An_{65-70} rarely ranging to bytownite, An_{78}) and zoned to sodic margins. Interstitial alkali-feldspar occurs rarely. There is commonly plentiful interstitial chloritic matter, enclosing trains of opaque ore cubes, perhaps representing devitrified glass. Some coarser opaque ore is also present.

Amygdales are small and uncommon and may contain chabazite. In one example (NI 1299) a central translucent cone of thomsonite, which from its optical properties has a composition in the middle of the range for this species, is surrounded by milky white tobermorite (films X 2676A, 2675A). In amygdales up to about 10 mm across in one specimen (NI 1615) a white rim is composed of gismondine,

around a cream-coloured or brownish core of tacharanite (X 4345, 4346). The margin of the tacharanite against the gismondine is more deeply coloured. Identification of the gismondine was suggested by its X-ray pattern: this could formerly be confused with that of garronite but the optical properties are distinctive (Walker, 1962c). The Carneal mineral has $\alpha = 1.522$, and confirmation of its identity has

Table 5 Chemical analyses of olivine-dolerite

	1	A	B	C
SiO_2 (per cent)	47.62	47.55	45.26	48.0
Al_2O_3	17.39	16.18	18.48	16.7
Fe_2O_3	2.90	2.46	1.62	2.0
FeO	7.12	8.35	7.62	8.8
MgO	8.58	8.62	9.23	8.9
CaO	11.64	11.86	11.41	11.3
Na_2O	2.03	2.19	1.98	2.3
K_2O	0.20	0.35	0.10	0.4
$H_2O > 105°C$	0.62	0.80	2.40	—
$H_2O < 105°C$	0.88	0.42	0.93	—
TiO_2	1.00	1.11	0.83	1.2
P_2O_5	0.15	0.14	0.10	0.1
MnO	0.17	0.16	0.16	0.3
CO_2	n.d.	0.03	—	—
Allowance for minor constituents	0.14	—	—	—
Total	100.44	100.24†	100.12	100.0
Ba* (mg/kg)	60	—	—	—
Co*	35	—	—	—
Cr*	<10	—	—	—
Cu*	140	—	—	—
Ga*	25	—	—	—
Li	n.d.	—	—	—
Ni*	200	—	—	—
Sr*	100	—	—	—
V*	250	—	—	—
Zr*	10	—	—	—
B	5	—	—	—
F	140	—	—	—
S	tr	tr	—	—
Norms				
Or	1.18	2.2	0.56	2.22
Ab	17.18	18.3	16.77	19.39
An	37.75	33.4	41.14	34.19
Di	15.34	20.0	11.93	17.37
Hy	14.62	7.1	4.55	7.01
Ol	6.28	11.9	17.66	14.39
Mt	4.20	3.5	2.32	3.02
Il	1.90	2.1	1.52	2.28
Ap	0.35	0.3	(0.25)	0.25

tr = trace; n.d. = not detected; *spectrographic determination; † BaO = 0.02.

1. Olivine-dolerite. Top of plug in centre of intrusion, Carneal, County Antrim. 1-in 29, 6-in 46 NE. NI 1605. Lab. No. 1966. (Analyst: G. A. Sergeant; Spectrographic work: C. Park.)
A. Olivine-dolerite, Scawt Hill. (Tilley and Harwood, 1931, p. 448.)
B. Olivine-dolerite bounding segregation vein, Scawt Hill. (Tilley, 1952, p. 531.)
C. Olivine-dolerite. Average of analyses of three plugs: Ballygalley, Ballymoney and Scawt Hill. (Patterson and Swaine, 1957, p. 325.)

been provided by reliable X-ray data that have recently become available (Berry, 1970).

A tachylitic olivine-dolerite (NI 1577) may be a chilled example of the normal igneous rock. It is composed of phenocrysts of idiomorphic olivine commonly 1.5 mm across and tablets of basic plagioclase commonly up to 1.5 mm long in a groundmass of pyroxene granules (pale augite) commonly 0.05 mm across, finer-grained labradorite, idiomorphic ore mineral crystals mainly pyrite, and interstitial pale dull purple glass. A brecciated olivine-basalt (NI 1599) contains dark grey angular fragments, in places more than 10 mm across, cemented by coarse calcite and analcime (X 4790). The feldspar of the olivine-basalt is partly chloritised. Some of the pyroxene spreads into the cementing veins as slender brownish prisms so that it is likely that the breccia is paulopost or contemporaneous.

In an olivine-basalt, probably chilled (NI 1579), amygdales are filled by a radiating acicular mineral probably natrolite, cemented by a more massive mineral probably chabazite.

Pyroxene-rich dolerite, gabbro, and basalt

With increase in the amount of pyroxene a distinction of rock-type must be made. On grain size the pyroxene-rich rocks are to be classed as dolerite, or less commonly gabbro, and rarely as basalt. They are typically hybrid types, containing calcium-silicate minerals, and showing characters intermediate between the dolerites and hybrid pyroxenites described below. The typical rock contains strongly zoned idiomorphic pyroxene (sahlite), which may reach several millimetres across, with laths of labradorite zoned to andesine. The most basic feldspar found was bytownite, An_{80}. There are clots of fine-grained thomsonite, plentiful robust ore crystals, some moulded on pyroxene, ilmenite rimmed by sphene, and discrete euhedra of sphene. There is a little alkali feldspar, mainly albite-oligoclase mantling labradorite and present also in coarse plates of late crystallisation. A little calcite is present and thomsonite is common. The sahlite has deeply coloured margins. The transition to more pronouncedly hybrid rocks is shown by the development of melilite and wollastonite, and in these examples a little aegirine or aegirine-augite occurs. Scarce olivine is iron-rich. Some interstitial green alkaline material may be devitrified glass. Some amygdales are present, and may contain calcite, thomsonite, and tobermorite.

Pyroxenite

This is a variable group of rocks that are characterised by coarse pyroxene, commonly sahlite, with wollastonite, melilite, labradorite, ore (commonly pyrrhotine), and interstitial alteration products. They are typically black, and rare amygdales are present. The pyroxene generally consists of stout, pale greenish sahlite prisms up to about 10 mm long, commonly with more deeply coloured margins. Some poikiloblastically encloses other minerals. In some examples melilite encloses and is moulded upon sahlite. In a fine-grained example (NI 1575) a small xenolithic area of bredigite is present. Sphene is present in scarce idiomorphic crystals; olivine is absent. The interstitial material is commonly thomsonite and alkali feldspar, aegirine, and calcite also occur. Some of the amygdales are of irregular shape, extending into the groundmass. They contain thomsonite, allophane and tobermorite, and tacharanite also occurs.

Titanaugite-melilite rocks

In these very rare rocks, titanaugite prisms reach 10 mm in length and may be in coarse intergrowth with melilite. Some plagioclase is enclosed by the titanaugite. The melilite (humboldtilite) forms hypidiomorphic tablets commonly about 5 to 10 mm long; some alters along cracks to an aggregate of fine-grained minerals, some fibrous, probably including cebollite, and some alteration appears to be thomsonite. Rare wollastonite prisms are present. There are tablets of basic plagioclase (bytownite, An_{85}, to labradorite), altered along cracks to thomsonite. Opaque ore minerals include pyrite, pyrrhotine and magnetite. Plentiful interstitial matter is commonly fibroradiate thomsonite, and patches of alkaline residuum contain fine-grained aegirine and probably also nepheline and its alteration products. Amygdales exceptionally reach 10 mm in length, mainly containing tobermorite.

Another scarce rock type of the endomorphic suite is melilite-rock (gehlenitic).

Exomorphic rocks

The metamorphism and metasomatism of the chalk has produced a number of calcium silicate rocks, in which there are rare carbonate-silicates; and metamorphism of flint has produced quartzites and wollastonite, plagioclase, xonotlite, and related assemblages.

Examples of larnite-magnetite rock are pale to dark grey and may have a dusting of portlandite on joint surfaces. Other minerals present include spurrite, scawtite, other calcium silicates, hibschite and perovskite. A ferrian spinel from a larnite-spinel rock (NI 1608) has already been described (Sabine, 1968). The melilite in larnite-magnetite-rock is one of the most highly gehlenitic melilites ever recorded, with Ge_{94} $Åk_6$ mol. per cent (Sabine and others, 1982). These rocks represent metamorphosed chalk and metasomatism shows clear effects in some.

Several examples of scawtite-xonotlite rock occur, probably representing metasomatised flint. These rocks are white or grey and look somewhat like flint. They are now composed of xonotlite containing spherulitic nodules of scawtite, with narrow veinlets of calcite and tobermorite. Other minerals present include hydrogrossular, spinel, melilite, larnite and merwinite.

A nodule, probably originally flint (NI 1305) is now an aggregate of quartz crystals 0.05 mm or less across. It is set in a grey hydrogrossular (hibschite)-calcite-spinel rock representing metamorphosed chalk. This rock is composed of abundant idiomorphic crystals or irregular aggregates of hydrogrossular in a groundmass of lowly refringent isotropic or weakly birefringent brown tobermorite. There are aggregates of calcite ($\omega = 1.658$) mainly less than 0.05 mm across, possibly with some carbonate of higher refractive index, scattered plagioclase in irregular crystals, commonly about 0.12 mm across, abundant euhedra of magnetite also commonly 0.12 mm across, and pseudomorphs in lowly refringent fibrous or platy minerals after a coarse prismatic mineral. The hibschite is of variable grain size but some crystals are about 0.05 mm across. There are abundant highly refringent granules commonly 0.02 mm or less across, probably perovskite. The rock is veined by drusy calcite with interstitial and centrally occurring isotropic tobermorite. This rock has much of the appearance of a tuff.

GENESIS

Unlike Scawt Hill, the Carneal rocks show no well-exposed direct contact between igneous rock and limestone that crops out, so that the genetical relationships must be inferred. However, the unusual chemistry and mineralogy of the rocks leaves little doubt about the major processes that have taken place. From the olivine-dolerite, assimilation of lime and silica leads to increase in diopsidic pyroxene, enrichment in iron in the olivine, and eventual disappearance of olivine and plagioclase. Hybrid pyroxenite is produced, which may have interstitial thomsonite. Melilite-titanaugite rocks, scarce among the Carneal assemblages, represent a further stage in which melilite has developed at the expense of pyroxene and plagioclase. Wollastonite may occur where there has been excess silica (sometimes provided by flint) and lime. An alkaline residuum is represented by assemblages with aegirine and nepheline, produced in small amount by continued assimilation of lime with desilication of plagioclase.

The pure limestone of the chalk has been metamorphosed and metasomatised to lime-silicates, at the highest stage of the prograde phase of metamorphism, principally represented by larnite rocks and related assemblages. Flint has recrystallised to quartzite or by metasomatism has been converted to wollastonite rock. The retrograde phase of metamorphism results in a suite of minerals mainly in order of increasing water content, including xonotlite, hydrogrossular, bicchulite, thomsonite, tobermorite, tacharanite and plombierite (Sabine and Young, 1975).   PAS

CHAPTER 12

Structure

GENERAL ACCOUNT

Very different tectonic styles characterise the rocks to either side of Belfast Lough. To the south in County Down the Lower Palaeozoic rocks are intensely folded and faulted, while to the north in County Antrim, the Mesozoic and Tertiary strata are only slightly folded although faults are prominent. The boundary between these very different tectonic regimes is not exposed but as the area lies approximately on strike with the Southern Uplands Fault of Scotland, it has been postulated that this fault extends into Ireland and may separate these structural regimes, as it does in some places in Scotland. The evidence relating to the possible extension of the Southern Uplands Fault into this area will be reviewed later. Details of the minor structures are contained in the chapters on each system.

The sequence of tectonic events which can be distinguished within the area of this sheet are:

1 Caledonian folding and faulting
2 Variscan (Hercynian) movements
3 Pre-Cretaceous movements
4 Intra-Cretaceous movements
5 Tertiary folding and faulting

CALEDONIAN FOLDING AND FAULTING

Although examples of actual fold hinges are comparatively rare, it can be demonstrated, from rapidly alternating younging directions that there is extensive folding throughout the Lower Palaeozoic succession. The most impressive display of folding occurs in the continuous shore sections around the Copeland Light House and Mew Islands and to the west of Grey Point. In these areas the folds occur in plan on the wave-cut rock platform and in elevation in the vertical walls of numerous sea-gullies eroded along fault lines.

Even more striking than the folding, and obvious during even a cursory examination of the Ordovician and Silurian rocks to the north and south of the Orlock Bridge Fault, is the highly 'tectonised' aspect of the Ordovician rocks. The cause of this difference in the degree of tectonisation is not known but could be due to:

1 a pre-Silurian period of folding and cleavage
2 differences in lithology allowing cleavage imprint to be developed more easily in Ordovician than in Silurian rocks
3 difference in depth of burial in the Caledonian tectogene

The occurrence of a pre-Silurian period of folding would be difficult to substantiate, as not only is there only a small stratigraphical gap in the succession through from Ordovician to Silurian at Coalpit Bay, but there is no discernible difference in the parameters of the structures in the Ordovician and Silurian outcrops. If difference in the depth of burial is invoked to account for the phenomenon then it would be reasonable to assume that the change in degree of tectonisation would decrease progressively upwards but there is no sign of such a gradual change; rather the difference is sharp and occurs at the Orlock Bridge Fault and is also conspicuous at Coalpit Bay, Donaghadee, where the Ordovician strata are intensely folded, cleaved and shattered and yet the adjacent Silurian mudstones are uncleaved and more gently folded.

In an attempt to account for the differences in aspect of the tectonic character of the Ordovician and Silurian rocks the parameters of folds, cleavages, tension and shear joints and faults were systematically measured and analysed. Three main phases of deformation were identified affecting both Ordovician and Silurian rocks:

Phase I An essentially periclinal swarm of isoclinal folds (F_1), with axes trending north-east to south-west and an associated axial plane (S_1) cleavage was developed in a stress field in which the maximum principal stress was aligned approximately north-west to south-east. Strike and wrench faults and lamprophyre dykes are associated with a later episode of this phase.
Phase II A series of small folds (F_2) with an associated axial plane cleavage (S_2) dipping at about 30° towards the north-east developed in a stress field in which the principal axis of stress plunged at about 70° towards the south-west.
Phase III A weak stress field in which the maximum principal stress was aligned east-north-east and produced a few small cross folds with axes trending north-north-west.

The age of these various phases of stress and the relationship to the major episodes of deformation which affected the British Isles, as enumerated by E. M. Anderson (1951), cannot always be directly established. However, the earliest epoch (Phase I) which produced the main folding, affects Upper Llandovery strata in the south of the Ards Peninsula (One-Inch Sheet 37), but does not affect the Lower Carboniferous rocks at Cultra. By analogy with the Girvan area of Ayrshire (Williams, 1959), this main fold phase is post-low Wenlock and pre-Lower Old Red Sandstone, and took place during later Salopian and possibly Downtonian times and is thus part of the Caledonian orogeny.

From similar reasoning the Phase II deformation, which affects the S_1 axial plane cleavage but does not affect the Lower Carboniferous at Cultra, is also part of the Caledonian phase of earth movements.

The Phase III cross-folds are also considered to be of Caledonian age and it is noteworthy that they are of similar styles and orientation to Simpson's (1963, p. 391) 'Third movement-phase (F_3)' folds in the Manx Slate Series of the Isle of Man. However, basalt dykes are aligned along the axial planes of small folds with axes trending north-north-west in the Triassic rocks at Holywood and Cultra and these are regarded as Tertiary structures. As these Tertiary folds are exactly similar in orientation and style to the small cross-folds in the Ordovician rocks, the Phase III cross-folds could conceivably be of similar age.

First Movement Phase (F_1)

Main Folds The first folds (F_1) are either slightly asymmetrical (Plate 19) or isoclinal concentric. The asymmetrical folds occur in massive greywackes interleaved with thin mudstones so that differential slip took place along the incompetent layers, while true isoclinal folds are confined to strata where argillaceous partings are more common.

The axes of these folds, are essentially periclinal and trend between 66° and 85°.

On Mew Island and Light House Island there are numerous first folds with features typical of concentric folding. On Light House Island the F_1 folds can be seen to be periclinal, although the modal fold based on bedding plane measurements consists of one limb dipping at 80° towards 327° and the other at 60° towards 165° while the fold axis plunges at up to 20° towards 242°. In the axes of the folds the well rounded turnover of the beds is often visible and as the plunge is usually low, the strike of the beds remains fairly constant although the dip changes.

The most characteristic feature of concentric folding in this area is the facility with which differential slip occurred along the incompetent layers and the consequent development of associated minor structures. This differential slip took place over a prolonged period of time ranging from the inception of folding to past the stabilisation of the regional cleavage which is commonly slightly sigmoidally flexured.

Between competent members of the series differential slip occasionally produced bedding plane slickensides with lineations orientated normal to the fold axes. At the same stage, particularly over the axes of folds, concentric shear planes developed and later were infilled to form lenticles of quartz or, less commonly, calcite. In addition, the crestal zones of anticlines are often irregularly net veined with quartz and, very occasionally, there are longitudinal quartz-filled gashes. On Light House Island, in the crest of an anticline, thin quartz veins occur in several beds of massive sandstone while in the shale layers between there are only occasional stringers of quartz—the incompetent beds having continuously yielded to pressure without the development of tension cracks.

There are many small strike faults dipping to north and south, usually concentrated near the hinges of folds but the most conspicuous are a series of northerly dipping reverse faults dislocating the northern limbs of folds. Particularly good examples of these can be seen in the Ordovician rocks in the north of Copeland Island where they also generated some drag folds.

Minor Folds Due to the relative movement along bedding planes separating competent and incompetent strata during the F_1 fold phase, two groups of minor folds with axes parallel to the F_1 fold axes developed—monoclines and sigmoidal flexures of the regional cleavage.

Before the imprint of the regional cleavage was finally stabilised on the folded strata several monoclinal flexures developed as a consequence of differential slip on bedding planes. These monoclines are intimately associated with the main folding as their axes are parallel or subparallel to those of the F_1 folds; however, the axial plane cleavage of the main folds cuts across the monoclinal folds which therefore precede the cleavage. The best developed of these monoclines lies on the northern limb of an overturned anticline on the foreshore between Luke's Point and Bangor Harbour where, in the lower 1·m of a rock knoll the beds dip 65° towards 164°; above 1 m there is a sharp kink and the beds dip at about 80° northwards. On both limbs of this monoclinal kink the regional cleavage dips 60° towards 170°. Consequently, this monocline must have been formed before the axial plane cleavage.

Since this monocline lies on the northern limb of an anticline overturned towards the north, differential slip should result in the beds further north moving upwards with respect to those nearer the core of the fold and that this happened is demonstrated by the generation of the above monocline.

Differential slip on the bedding planes continued until after the stabilisation of the axial plane cleavage (S_1) and, in fact, twisted the cleavage giving rise to distinctive sigmoidally curved cleavages. These folded cleavages typically occur in the more highly cleaved silty mudstones or other incompetent beds lying between competent strata of the Ordovician series.

On Light House Island the relationship of these sigmoidal cleavages to the axis of an anticline is clear and the sense of differential slip on the bedding planes is that normally to be expected in a concentric anticline; that is, the upper competent layers moved towards the axis of the fold with respect to the lower layers (Figure 24).

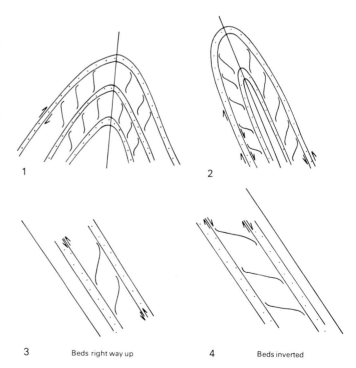

Figure 24 Sigmoidally flexured cleavages in incompetent strata

Many isolated examples of these sigmoidally folded cleavages occur on Light House Island and Copeland Island and also on the coastal section between Orlock Head and Orlock Bridge.

A further group of minor folds for which differential slip is responsible is only poorly represented, mainly on the coast section within 140 m N of Orlock Bridge Fault. Here, alternating siltstones and shales, cleaved coincidentally with the bedding, are folded into a series of small concentric anticlines and synclines with axes plunging 30° towards 76°. These folds are drag folds produced in a series of beds which are incompetent with respect to the massive sandstones in which they are sandwiched.

Cleavage The axial plane cleavage (S_1) associated with the F_1 folds is especially well developed in the Ordovician and less commonly in the Silurian rocks. Both slaty and fracture cleavages occur in the Ordovician series but in the Silurian strata generally only fracture cleavage is developed. These cleavages are intimately associated with the main fold phase and are parallel or subparallel to the axial planes of the folds and generally dip at around 80° SE, except in the fold hinges where they diverge markedly, and are disposed in a fan converging towards the cores of anticlines. Also, on the Copeland Islands the cleavage dips steeply north-north-west.

In the cliff section to the south of Orlock Head and in Light House Island the fracture cleavage in the argillaceous layers is often sigmoidally twisted as it approaches the arenaceous beds. As these twisted fracture cleavages are sometimes replaced at the bedding plane separating the competent and incompetent layers by concentric shear, the sigmoidal twisting of the cleavages is probably due to drag rather than change in lithology. De Sitter (1956) notes that the sharp bedding planes separating competent from incompetent layers often act as concentric shear planes while fracture cleavage is developed in the incompetent beds. Changes in orientation of cleavage caused by its passing from one lithology to another are often pronounced, particularly in areas where slaty cleavage predominates and passes from a mudstone into a siltstone. The sigmoidal cleavages cannot, however, be attributed to change in lithology as the twist occurs within a homogeneous layer. Thus, the sigmoidal cleavage is associated with the relative drag movement produced during concentric shear.

Under these conditions of generation the sense of the sigmoidal twist is always the same on a particular limb of a fold but changes across the hinge. In the few cases on Light House Island where the position of these sigmoidal cleavages on the limb of a fold and their relative orientation to the fold axis is known, the sense of the movement producing the twist is always the same. That is, the outer layer always moves upwards towards the hinge of the anticline.

Second Movement Phase (F₂)

In the F_2 stress field a series of small obtuse-angled folds were developed affecting the S_1 cleavage. The fold hinges are angular and the axial planes dip at 10° to 25° NE and are always marked by a well developed strain-slip cleavage (S_2).

Typically the F_2 folds occur in the almost vertical, highly cleaved shales and grits on Ballymacormick Point; in the area west of Bangor Bay and at various places along the shore section between Smelt Mill Bay and Craigavad. In these areas the intersection of the second axial plane cleavage (S_2) with the First fold cleavage (S_1) produces strong, nearly horizontal, 'B' lineations. The vast majority of all the F_2 folds are disposed in the same attitude and, when viewed in elevation, the short limb always dips steeply north.

Third Fold Phase (F₃)

The final phase of folding is poorly represented by a few small cross-folds on the shore at Orlock Head, Carnalea Golf Club and near Craigdarragh. These are gentle, open symmetrical folds with vertical axial planes and axes plunging by the same amount as the dip of the main fold limbs on which they occur towards 350°. There is no direct evidence of their age and it could be argued that they are, in fact, due to pre-Cretaceous movements.

Joints

Joints are extensively developed in all the Lower Palaeozoic rocks from Holywood to Kinnegar Point. In shale and mudstone layers they are often quite irregular but in all beds of coarser than silt grade material they are flat surfaces, commonly with veins of quartz or calcite. As has already been mentioned the attitude of fold axes and the inclination of bedding planes throughout the area is not constant and the possibility of this disparity in trends being due to block rotation within faulted boundaries has been considered.

If the block rotation took place after the development of jointing associated with the F_1 fold phase then it should also be discernible in a corresponding swing of the a–c tension joints. To test this hypothesis more than 3500 joints were measured and the results were plotted as stereograms.

To the west of Bangor where the modal fold axis plunges 30° towards 248° there is a set of joints dipping 60° towards 245° which is within 3° of being normal to the fold axis. Similarly, in the Conlig Lead Mines area where the strike of the beds is between 65° and 70° there is a set of near vertical joints striking 161° and on Ballymacormick Point 159° trending joints dip 80° west and the beds strike 65°. In contrast to these areas between Orlock Bridge and Foreland Point the axes of periclinal folds trend between 74° and 85° and the associated a–c joints strike 172° and dip east at 70° so that it could be argued that the whole block has been tilted towards the west and rotated through about 10° in a clockwise direction with respect to the areas described previously.

From this it was concluded that where the fold axes strike east–west the associated a–c joints are more nearly north–south than in areas where the fold axes trend east-north-east. However, on both Copeland Island, where the beds strike 85°, and on Kinnegar Rocks, where the strike is 65°, the a–c joints strike 170° although their inclination is different.

Plate 19 Asymmetrical anticline in massive greywacke and thin shale. Well developed axial-plane cleavage. Ballykeel Quarry   [NI 426]

Plate 20 Post-cleavage dome in massive mudstones, cleaved coincidentally with bedding. Foreshore due east of Galloways Bridge   [NI 420]

In addition to the main fold phase a–c joints there are two other groups of joints which do not always appear prominently on the stereograms but which are remarkably constant throughout the region. These are:

(a) a group of low angle quartz-veined joints dipping at 40° towards 30° which are essentially parallel to the axial planes of the F_2 folds. These joints are only abundant from Craigavad to Ballymacormick Point. However, they also occur on Orlock Point and in the Silurian rocks north of Donaghadee. In the Orlock Point area they dip at 45° towards 45°.

(b) an almost vertical set of joints trending about 10° to 20° and on which there is often calcite veining. This set is sporadically developed on Kinnegar rocks and less commonly in the Conlig Lead Mines area and in the central block of the Copeland Island. In the first two areas the strike of the beds is 65° and in the third 90°; thus, there is no relationship between these joints and the fold trends and the joints must have been formed after the block rotation and are possibly late-Caledonian or even Armorican in age.

Faults

As has already been briefly mentioned there is faulting in the region but, due to the uniformity of lithology and lack of any distinctive marker horizons, actual fault dislocation amounting to more than a few centimetres is usually not immediately obvious, though the trend of the fault may be.

Of the large Caledonian faults the most important are named on the published one-inch map and are:

1   The Orlock Bridge Fault, trending 70° is a normal fault downthrowing to the south, possibly with a slight hinge movement so that the amount of dislocation is greater on Copeland Island than at Conlig;
2   Copeland Island Channel Fault, trending 130° has a dextral wrench component of movement which dislocates the Orlock Bridge Fault by 300 to 400 m and lies somewhere in the channel between Copeland Island and the mainland.
3   Orlock Tunnel Fault, trending 30° is probably a sinistral wrench fault separating blocks in which the trend of the fold axes is markedly dissimilar;
4   Garrahan Isle Fault, trending 15° may be a sinistral wrench fault but there is no evidence other than its trend. It does, however, separate two of the 'blocks';
5   Ballyholme Fault Complex is a series of 12° trending faults seen in the rocky promontories on both sides of the bay and which also produce features inland. They are sinistral wrench faults which are postulated to offset the westward extension of the Orlock Bridge Fault by about 350 m.

There are numerous small faults and slickensided surfaces which indicate the relative displacement along them and on the assumption that these minor faults originated in the same stress field as the major faults some 245 surfaces with known directions of movement were measured and the results plotted on rose diagrams (Figure 25). 'A' is plotted from wrench faults from areas where the main cleavage trends approximately 65° and 'B' from faults from areas where the strike is approximately 85°.

In 'A' the two most prominent maxima are between 140° and 150° and between 10° and 30°. The former maxima is made up of dextral wrench faults and the latter sinistral wrenches.

In 'B' the largest maxima are between 10° and 40° and between 130° and 140°. In addition there are many dextral faults trending more north-easterly than in 'A'. The 10° to 40° maxima is made up of sinistral wrench faults and a 130° to 140° group of dextral wrench faults.

It will be noted that several trends on which major faults occur do not form prominent maxima; this is due to the fact that these diagrams are quantitative evaluations of the various trends of fault planes and maxima need not bear any relationship to the relative importance of the group of faults which they represent and, in fact, it was shown in the Southern Highlands of Scotland by Smith (1961, p. 150) that the expected complementary wrench faults to a set of well developed major NE-trending wrench faults were represented by a series of minor faults which were not large enough to be represented on a geological map, but were strongly represented in the stereographic plot of several hundred fault surfaces. For example, there is a considerable number of small dextral wrench faults trending 10° to 20° which can be related to the F_2 stress field but these faults are not known to produce any spectacular dislocations.

In addition to these minor faults of known type many sea-gullies and bays have been eroded along fault lines. These gullies are sometimes floored by shattered material and the breccia is often reddened by secondary hematitic and limonitic staining.

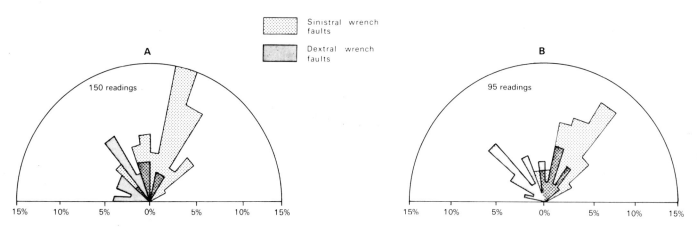

Figure 25   Rose diagrams of wrench faults in areas where: (a) fold axes trend 65° and (b) fold axes trend 85°

The Southern Uplands Fault   It has been postulated (George, 1953, 1960; Charlesworth, 1953, 1963; McKerrow 1959, Leake 1963, Anderson 1965) that the Southern Uplands Fault, which is such an important feature of Scottish geology, extends south westwards into Ireland. The exact position of the fault as Anderson (1965, p. 383) stresses, however, has never been established and its extension is postulated mostly on the grounds of outcrop distribution as shown on small scale maps and in particular on the apparent straightness of the north-western margin of the Down–Longford Lower Palaeozoic outcrop.

Manning and others (1970, p. 120) have emphasised that the linearity of this boundary is illusory and that in the Belfast area (One-inch Sheet 36) it is markedly sinuous, reflecting the low relief of the Lower Palaeozoic rocks and the unconformable nature of the base of the Permo-Triassic sediments and that only in the Castlereagh area does the boundary correspond to a pronounced pre-Permo-Triassic topographic feature. The geophysical evidence also indicates that there is no sub-Permo-Triassic feature which could be attributed to the extension of the Southern Uplands Fault close to the margin of the Lower Palaeozoic outcrop and that the basement to the Mesozoic sediments shelves gently towards the north-west, as was confirmed by a borehole at Long Kesh (Manning and others, 1970, p. 204) which encountered Lower Palaeozoic rocks beneath a thin (70 m) Permo-Triassic cover.

George (1960, p. 83) suggested that 'the Southern Uplands Fault may be precisely located at Cultra, where the Carboniferous rocks are broken by a fracture that brings together Permian rocks overlying Dinantian on one flank; Bunter overlying Ordovician on the other'. It would appear that in this passage George is referring to the faults affecting the Carboniferous, Permian and Triassic strata which are probably of Tertiary age. It is improbable that any one of these is the Southern Uplands Fault as George suggests.

It is possible that the fault bringing the Carboniferous and Ordovician rocks in Cultra Glen into juxtaposition may be the Southern Uplands Fault as it satisfies the necessary requirements of age and direction. However, the available field evidence does not indicate an unusually large fault and such a correlation is thought to be unlikely.

On the basis of the regional gravity map of Northern Ireland, Bullerwell (1961b, p. 256) argued that the Southern Uplands Fault could not be identified but Anderson (1965, p. 388) postulated that the considerable magnetic gradient located across the fault in the south-west of Scotland (Aeromagnetic Map of Great Britain and Northern Ireland, 1964) could be traced across the North Channel to Whitehead and tentatively suggested that, if the fault existed, then this was the most likely line. However, it is pointed out in Chapter 15 (p. 89) that the anomaly noted in the south-west of Scotland probably has nothing to do with the Southern Uplands Fault but reflects the existence of ultrabasic rocks at depth. Furthermore, in the south-west of Scotland the effect of the fault is reduced and George (1960, p. 39) notes that 'its apparent throw is diminished, perhaps reversed, in Galloway . . .'.

It has been postulated by Dewey (1971, p. 219) that in Ordovician and Silurian times this area lay astride a NE-trending trench associated with a northward dipping and descending oceanic plate so that while oceanic plate was being consumed in the trench, sediment from the rising Highlands area to the north was accumulating both in the trench and on the adjacent plates. This sediment built up a thick pile of flysch which, beginning in late Upper Llandovery times, was progressively deformed and welded to form a growing land area (Cockburnland). By early Devonian times a land mass covered the Southern Uplands–Down–Longford area and sediment was being supplied to the Midland Valley from both the metamorphic Highlands to the north and the lower grade Lower Palaeozoic rocks of Cockburnland in the south. In Scotland during the Devonian era the Midland Valley was, at least in places, a fault-bounded graben but in north-east Ireland there is no evidence, at present, that this basin of sedimentation existed at all in Devonian times. At Cultra no Devonian rocks are exposed below the Carboniferous (Dinantian) beds and the Langford Lodge Borehole (Sheet 36) passed directly from Permo-Triassic rocks into Lower Palaeozoic grits. If the Southern Uplands Fault extends into Ireland and has any importance in controlling Upper Palaeozoic sedimentation, then it probably lies somewhere north of Carrickfergus. However, by early Carboniferous times the margin of the basin must have been overstepped and the old Red Sandstone strata were overlapped, so that the Dinantian beds at Cultra rest unconformably on Ordovician.   AEG

VARISCAN MOVEMENTS

The Variscan (Hercynian) orogeny, which in Southern Britain and Ireland caused widespread folding on an east–west axis, caused stresses in the northern areas of the British Isles which renewed movements along the Caledonoid structures. In the Central Valley of Scotland these movements affected Carboniferous sedimentation from mid-Dinantian times and are reflected in the lateral variations of thickness which now give the Ayrshire, Lanark–Stirling and Fife–Midlothian coalfields.

In this area there is little direct evidence that these processes were continued south-westwards into Ulster, but Wright's (1919, 1925) postulation of a sedimentary basin in the north-eastern part of the sheet has been supported by the results of gravity surveys which indicate a major sedimentary basin, partly explored by the Larne Borehole (Manning and Wilson, 1975).

Reactivated movement on Caledonian faults in the district, and possibly fresh faulting, must have been on a considerable scale. Massive brockram of Permian age, known from boreholes at Scrabo (Newtownards), Avoniel (Belfast) and elsewhere, must have been derived from steep slopes, probably controlled by major faults.

The fault, referred to earlier, which cuts out the lowest part of the Carboniferous succession in Cultra Glen may form the boundary between the Carboniferous and Ordovician rocks for some distance but has not been so shown on the one-inch map because it is conjectural and to draw it on would invite unjustified correlation with the Southern Uplands Fault.

Lead-zinc mineralisation at Conlig and, on a smaller

scale, at Cultra and Craigavad, is probably of Hercynian age.

PRE-CRETACEOUS MOVEMENTS

From the record of minor non-sequences and uplifts during the Cretaceous it is clear that there was some crustal instability during much of that era, but the sporadic occurrence of the Lias and Penarth Group rocks beneath the Hibernian Greensands suggests that there was some pre-Senonian warping which folded the beds and facilitated their partial erosion. To this warping may be ascribed small-scale folding in the Triassic rocks.

On the coast between Carrickfergus and the western edge of the sheet the Mercia Mudstone Group is affected by a series of rolling, open folds ranging in amplitude from a few metres up to 200 m. These folds plunge mostly to the north-west at low angles which are difficult to determine but which are probably near to that of the regional dip. About 1 km SW of Greenisland the Sherwood Sandstone Group is brought in by a north-easterly trending periclinal fold with an amplitude of about 100 m. Several smaller periclinal folds, with axes parallel to the major folds, also occur in this section.

It is noteworthy that on this well exposed coast section no faults are seen and it is thought that this may be because of the incompetence of the rocks of the Mercia Mudstone Group which have deformed by flexure rather than fracture.

On the south side of the Lough, in the outcrop of Triassic and Upper Palaeozoic rocks, there is a considerable amount of folding and faulting. The faults disrupt all the major rock units present but the folding is very unevenly distributed, affecting some beds more than others.

INTRA-CRETACEOUS MOVEMENTS

The depositional history of the Cretaceous rocks in County Antrim indicates continued crustal instability during much of Cretaceous time. There is abundant evidence of epeirogenic movement, warping and faulting in adjacent districts (Reid, 1971, p. 125; Manning and others, 1970, p. 120) but probably because of the indifferent exposure, the only clear indication of these processes is in the intraformational non-sequences noted at Cloghfin.

As the highest beds of the Ulster White Limestone known in Antrim—the Maastrichtian—show the same signs of post-depositional erosion controlled by faulting, it is clear that the earth movements continued into the post-Maastrichtian period and probably merged into the convulsions which marked the onset of early Tertiary vulcanicity.

TERTIARY FOLDING AND FAULTING

The onset of extrusive volcanic activity in the Palaeocene was undoubtedly accompanied by faulting. Evidence for this can be seen in north Antrim (Wilson and Robbie, 1966, p. 252), but which is lacking in the Carrickfergus district. Movement on some of the many faults which affect the lava pile may have begun at an early stage and have involved the reactivation of earlier Caledonian fractures. The regional downwarping of the Lough Neagh basin, which gives the whole Antrim plateau its gently inward dip, has affected the lavas and Mesozoic rocks north of the Lough but most of the steeper dips are due to local fault-controlled tilting.

Two groups of faults of Tertiary age are recognised—a dominant series trending north-north-west and a subsidiary series trending north-east.

The NNW-trending faults

A series of subparallel faults affects the area north of Belfast Lough. The Woodburn, Carneal, Beltoy, Larne Lough and Black Cave faults have downthrows to the east, as have several smaller unnamed faults, while the Black Hill and Castle Dobbs faults downthrow to the west. The actual displacements of these faults are difficult to estimate because of the lack of stratigraphical markers in the lavas and poor exposure of the underlying sediments, but in most cases they appear to be of the order of 100 to 200 m. The Carneal Fault, which may well extend south to appear as the fault at Milebush, and the Larne Lough Fault appear to have significantly greater throws than the others.

On the south side of the Lough the Upper Palaeozoic and Mesozoic rocks are disrupted by a series of faults trending in a similar direction to the major faults on the north side of the Lough. In contrast, however, these faults with one exception, have a consistent downthrow to the west. A minimum downthrow of 100 m is indicated and it may well be as much as 200 m. These faults have the effect of progressively bringing in younger strata to the west—Carboniferous strata are brought against Ordovician, then Permian against Carboniferous, and finally Triassic against Carboniferous.

The most extensive of these faults is the Craigantlet Fault which brings Carboniferous sediments against low-grade metamorphic Ordovician rocks on the coast whilst its extension inland is considered to displace the faulted Ordovician–Silurian boundary. This is the only fault of this generation which may be traced into the Lower Palaeozoic strata and along much of its length a substantial olivine-dolerite dyke is intruded.

From the similarity in trend between these faults and the faults on the north side of the Lough it is considered likely that they belong to the same generation.

The NE-trending faults

Few NE-trending faults cut the Lower Basalts, unlike the area to the west (Manning and others, 1970). Faults north of Castle Dobbs and south of Seamount affect the lavas and Cretaceous outcrops and may be associated with some gentle folding on an axis trending north-east. The fault, north of Castle Dobbs, appears to deflect the Castle Dobbs Fault, and is thus younger than several minor ENE-trending faults which offset the Carneal Fault. Another group of faults affect the Cretaceous outcrop between Bentra and Red Hall where small displacements could reflect the existence of an underlying dislocation such as the westward extension of the Southern Uplands Fault.

AEG, HEW

CHAPTER 13

Pleistocene

PRE-GLACIAL TOPOGRAPHY

The detailed information on depth to rockhead obtained from site investigations in the Belfast area allowed the production of a contour map of the pre-Glacial land surface (Geology of the Country around Belfast, Plate 9, and 3 inch to 1 mile map: Geology of Belfast and District). This shows that the pre-Glacial Lagan flowed in a narrow, steep-sided valley with a bottom level about 60 m below present sea level, and that tributary streams were also in narrow gorges, incised in the Triassic sandstones, which graded steeply towards the main channel.

Of those which can be plotted in the Belfast area, the course of only one can be indicated in this Sheet, a stream which flowed north from the Knock district and ran across the intake area west of Kinnegar. It may be assumed that the main buried channel is somewhere in the middle of Belfast Lough and that there may be a parallel gorge towards the northern shore. Where these two converge is not known. It is also probable that even quite minor streams—the Woodburn River, Kilroot River, Crawfords Burn—on both sides of the Lough may have cut deep-sided channels in the soft Triassic strata during that late-Tertiary or early-Quaternary period when the sea level stood 100 m or more below present level. The courses of these channels are quite unknown, but their possible existence may be of relevance when off-shore engineering works are contemplated.

GLACIAL DEPOSITS

Although rock is exposed, or covered only by thin soil, over about one-fifth of the land area, the greatest part of the district, is mantled by glacial drift, mainly boulder clay or ground moraine. Glacial sands and gravels occur on a limited scale on the County Down side of the lough.

The area was completely covered by ice-sheets for a large part of the Glacial period, but the ice had little effect on the topography, apart from scraping off the weathered surface, resulting from millions of years of Tertiary erosion and deposition, and replacing it by a veneer of ground-up rock debris—the boulder clay.

The earliest observations on the drift of this area were by Agassiz (1842) who mentioned striae at Donaghadee, and by MacAdam (1850a) who described sections along the railway to Holywood. Close (1866) described striae from Bangor, Groomsport and Donaghadee and ascribed them to ice moving from the north-north-east, which he assumed to have been deflected by ice from a north-north-west source. Of the two original memoirs which covered the area (Hull and others, 1871; Hull, 1876), the former gives a brief account of the Glacial deposits but the latter is very inadequate. The County Down area is partly covered by the

Special Memoir *The Geology of the Country around Belfast* (1904) which gave a comprehensive and detailed account of all the drift.

From 1894 onwards the Belfast Naturalists Field Club published a number of reports on the distribution of glacial erratics in the drift, some of which referred to this area. Two important papers on the glaciation of north-east Ireland by Dwerryhouse (1923) and Charlesworth (1939) set the area in its regional context. The former postulated the effects of Scottish ice, moving south-westwards across Antrim and Down; while the latter introduced the concept of a re-advance of this ice, after an interstadial period, which affected only the coast but produced profound disruption of the drainage and explained the distribution of fluvio-glacial features.

In recent years Synge and Stephens (1960, 1966), Prior (1966), Vernon (1966), Hill and Prior (1968) have re-assessed the evidence and conclude that although ice from Scotland had some effect in the last glaciation—the only one for which any evidence is preserved in this area—most of the drift deposits are the result of ice moving from an ice-shed trending north-eastwards from Lough Neagh across the North Channel. Hill and Prior (1968) claim that in north Down there are two boulder clays, varying in till fabric and also in colour, particle size analysis, carbonate content and erratics.

Charlesworth (1963, 1973) refused to accept these arguments on the grounds that the use of till fabrics in shallow drift is fraught with errors. However, on superficial examination, at least, two distinctive types of boulder clay were recognised and there are definitely two tills in the Lagan Valley to the west.

Boulder clay

Boulder clay, till, or ground moraine, is the ubiquitous product of the glaciation and covers four-fifths of the land area of the Sheet. It is mostly in the form of a relatively featureless spread, particularly on the low ground north of Belfast Lough but in County Down it locally forms rounded hills—drumlins—which may have rock cores in some cases. The elongated form of these hills, may indicate the direction of ice movement.

The characteristics of the till depend primarily on the underlying rock and, in spite of the presumed mode of formation beneath a moving ice sheet, carry-over of material from one outcrop to another is remarkably limited, though far-travelled erratic pebbles and blocks of chalk, flint, basalt, dolerite, granites, schists, Ailsa Craig micro-granite and quartz-porphyry are common. Where the underlying rock is soft it is difficult, in some localities, to distinguish between rock *in situ* and the disturbed material which grades up into the boulder clay. With massive bed-rock, basalt or greywacke, the junction is clear-cut but the lowest part of the till is heavily charged with blocks and pebbles of the underlying rock in a matrix of comminuted local debris. The proportion of local boulders decreases upwards and erratics become more abundant in the upper levels. Tills are subject to weathering and only in deep and fresh excavations are they seen in an unweathered condition. Tills from basalt are dark brown when fresh; those from the Lower Palaeozoic rocks are dark bluish grey, and only those derived from the Triassic rocks are reddish brown, although all look red-brown on weathering.

On the basalt plateau the boulder clay is usually fairly thin and is a reddish brown somewhat friable clay with a very heavy load of basalt blocks and few exotic pebbles, mainly flint.

The till on the Triassic outcrop is a tough, plastic clay with a variable proportion of boulders—mainly basalt—but with much chalk and flint, which can locally be dominant. Much of the basalt is derived from the numerous dykes in this area, and the matrix is reworked Mercia Mudstone, which gives a tenacious plastic clay, carrying no water and devoid of the sand lenses which occur in other areas.

On the County Down side of the Lough, though most of the outcrop is of Lower Palaeozoic shale and greywacke, there is some variation in the nature of the till from west to east. In the area south of Holywood-Cultra the clay shows signs of its derivation from the Mesozoic and Carboniferous rocks which underlie the coastal strip at the head of Belfast Lough, while east of Grey Point the clay is tough, blue-grey and compact. There is, however, some suggestion in both areas of a second red, sandy till, seen on the face of the hills above Holywood, in the coastal strip west of Grey Point, and also in the area around Donaghadee. Hill and Prior (1968) record a second till throughout the region but this cannot be confirmed by Geological Survey observations.

The thickest drift occurs on the lower ground to the east and north of the Craigantlet Hills and progressively thins up the sides of the hills. On the high ground it occurs only as pockets infilling depressions in the rock. On the lower ground the clay is, typically, moulded into subdued drumlinoid hills with axes trending just east of north but in a coastal strip, of varying width, extending from the southern boundary of the sheet to Ballywilliam, about

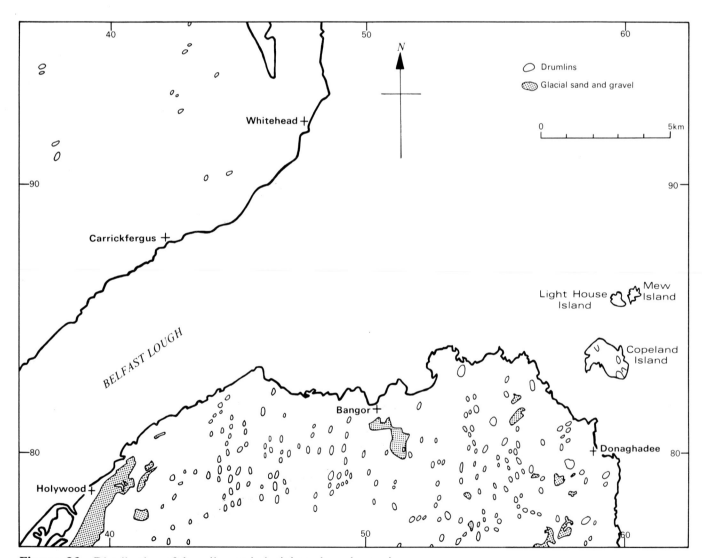

Figure 26 Distribution of drumlins and glacial sands and gravel

1.5 km N of Donaghadee, and in the area to the west of Grey Point, the drumlins trend more north-easterly (Figure 26) and are made up of distinctive bright-red boulder clay. In addition, in the vicinity of the western limit of the north-east trending drumlins in the Donaghadee area, there are numerous small patches of glacial sand and gravel indicating that these drumlins were formed near the margin of an ice-sheet, probably ice of the 'Scottish re-advance'.

Many drumlins are composed of boulder clay but in some cases they are rock cored. Irish Hill, south-west of Holywood, is solid rock with only a thin impersistent veneer of clay on the highest places.

There are few clear cut sections in the boulder clay but in those seen, mostly in the Craigavad area, there are wisps and lenses of sand and laminated clay within the mass of the till. During the construction of the Belfast–Bangor railway MacAdam (1850) observed many good sections in which bands and lenses of similar material occurred.

Glacial lakes and fluvio-glacial deposits

Peri-glacial sands, gravels and laminated clays are confined to the County Down area. They were deposited during the decay of the ice-sheets when outwash debris was carried by meltwaters and deposited around the margins of the stagnant ice, or in ice-ponded or sub-glacial temporary lakes. Such a lake was formed in the lower valley of the Lagan when Belfast Lough was blocked by ice, probably ice of the Scottish re-advance, and meltwaters carried sand and clay into the standing water of this Lake Lagan. The water level in the lake was controlled at various times by overflows through the Soldierstown gap into Lough Neagh and the Dundonald gap into Strangford Lough. Both are just under 30 m OD and the top of the Knocknagoney sands is also at this level. Further from the shore, laminated, plastic clays were laid down and have been found in boreholes in the harbour area.

While the low ground of the Lagan Valley and Belfast Lough was occupied by ice, marginal lakes formed in depressions along the northern side of the Craigantlet and Holywood Hills and overflowed by small channels to the south. A small patch of sand and gravel in Holywood glen and laminated clays at Ballysallagh are relics of these local and transient inundations.

On the County Antrim side there is also evidence that temporary ice-ponded lakes were formed in Woodburn Glen, which overflowed at various times across North Carn by a rock-cut spillway to the north, across the face of Knockagh by small steeply inclined 'chutes' and possibly an ice-margin channel along the face of the escarpment to the west. Probably at early and late stages in the last glaciation the drainage, flowing into Larne Lough, was obstructed by ice in the North Channel, and glacial lakes in Glenoe and in Larne Lough overflowed to the south, across the col above Beltoy and through Whitehead, respectively.

At a late stage in the decay of the Scottish Re-advance Ice, morainic sands and gravels were deposited around the margins of the ice on the low hummocky region between Bangor, Conlig and Donaghadee and are now exposed as small outcrops of poorly to well-sorted sand and gravel interspersed among the low drumlins.

Glacial striae

The harder rocks, particularly the greywackes, often show ice-moulded surfaces, on some of which it is possible to detect the scratch marks made by stones embedded in the ice. Few of these were seen in the recent resurvey but in the 1904 Memoir a considerable number are recorded, indicating ice movement over north Down, predominantly from the north but with a north-north-west trend in the extreme east. A few anomalous examples in the southern part of the area showed ice from the north-east, and one single example below high water in South Woodburn reservoir was orientated east–west. The dominant direction agrees well with the orientation of the few well developed drumlins. HEW

CHAPTER 14

Recent

With the melting of the ice-sheet, the position of the shore line in this area was affected by both the world-wide rise in absolute sea level and the isostatic rise of the formerly ice-loaded land surface. These factors, influencing the relative levels of land and sea, operated at different speeds, but when the two rates locally coincided for a period of time a wave-cut platform was formed.

Recent work (Stephens, 1957, 1958, 1963; Orme, 1966; Synge and Stephens, 1966) has shown that the older classification of these raised beaches into 25, 50 and 100 ft levels is an over-simplification. In particular, the well marked rock platform a few metres above present high water mark, hitherto described as the 'Twenty Five Foot Raised Beach', is mainly pre-glacial or intra-glacial in age, though it is locally covered with post-glacial beach material. Most of the coastal rock cliffs are coeval with the rock platform.

Higher platforms, sometimes with beach deposits, are late-Glacial in age, while the pre-Boreal period, when sea level fell below the present, allowed the formation of peat at levels which are now below high water mark. In the Lagan estuary, the Flandrian transgression swamped the peat and the Estuarine Clay was deposited in the brackish water tidal flats of the river mouth, which, as the transgression continued, deepened to give more saline conditions.

Recent deposits include present-day beaches and estuarine muds, and small areas of river alluvium. Land-slips below the basalt scarp have been moving intermittently from early post-glacial times to the present day.

ESTUARINE CLAY

This formation underlies the intake area on the south side of Belfast Lough, at least as far as Holywood and may also underlie part of the sea-floor north of the Victoria Channel. The main part of the outcrop is beneath Belfast Harbour and has been described, with fossil lists, in *The Geology of Belfast and the Lagan Valley* (Chapter 12, Appendix 4).

The Estuarine Clay has been divided into three members:

Upper Estuarine Clay	*Thracia convexa* Zone	Pollen Zone VIIa
Intermediate Estuarine Clay	Characterised by species of *Pholas*	Pollen Zone VIIa
Lower Estuarine Clay	*Scrobicularia plana* Zone	Pollen Zone VIb and c

The formation is known in the area of this Sheet only from borehole records and a few small sections seen at low water.

Site investigation bores for the oil refinery on the area of reclaimed land in the extreme south-west corner of the sheet, showed the base of the formation at depths ranging from 5 m to over 20 m. The area has been raised above sea level by dredged material pumped into polders and the bores showed several metres of fill and above one metre of recent silt and mud over the Estuarine Clays.

One of these boreholes was collected in detail, Estuarine Clay being present from a depth of 4 m. Above this, three samples from between 3 and 3.6 m yielded a fauna containing *Cerastoderma edule* and *Ostrea edulis*. These, together with the absence of *Abra alba* and *Corbula gibba*, suggest the present-day fauna.

Between depths of 4 and 10.7 m, the abundance of *Abra alba* and *Corbula gibba*, at most levels, indicates Upper Estuarine Clay (*Thracia convexa* Zone). Twelve species of gastropods and twelve species of bivalves were recorded as follows:

Summary of fauna from depths of 4 to 10.7 m (Upper Estuarine Clay)

Gastropods (12 spp.): *Alvania beanii calathus, Aporrhais pespelicani quadrifidus, Bittium reticulatum, Cingula vitrea, Gibbula cineraria, Haedropleura septangularis, Hydrobia ulvae, Littorina saxatilis, Nassarius pygmaeus, Ocenebra erinacea, Rissoa lilacina, R. membranacea.*

Bivalves (12 spp.): *Abra alba, Chlamys varia, Corbula gibba, Modiolus modiolus, Mysella bidentata, Mysia undata, Nucula sulcata, Ostrea edulis, Parvicardium exiguum, Saxicavella jeffreysi, Spisula sp., Venerupis rhomboides.*

Between 10.9 and 11.2 m only *Chlamys varia* and *Ostrea edulis* were collected. The abundance of *Ostrea* at this level suggests the Intermediate Estuarine Clay.

No Lower Estuarine Clay was identified on this site.

At Holywood Gasworks, Praeger (1892, p. 235) stated that excavations during construction in 1860 penetrated 22 ft (6.7 m) into the clay and that piles for a chimney went down 38 ft (11.6 m) before reaching firm ground. Welch (1914, pp. 239–240) recorded a section dug at a vertical retort house under construction at that time which revealed:

Section at Holywood Gasworks

	Ft	in	(m)
Topsoil with worked flints		6–8	(15–20 cm)
Clay, red and grey with chalk fragments	2	6	(0.75)
Clay, sandy, blue, with fragmentary shells	4	0	(1.2)
Clay, unctuous, blue, free from shells	28	0	(9)
Sand, hard grey with large oysters	2	0	(0.6)
Clay, hard red			

To the north-east, at the mouth of a stream passing Seapark Terrace, 0.6 m of peat could be observed between high and low tides filling a shallow trough in boulder clay. Between the peat and the boulder clay was 0.3 m of fine-grained red sand. Trunks and branches of Scotch fir, willow, oak and hazel were found with some hazel nuts.

Peat was also observed in the stream banks beneath stratified gravels of the raised beach.

In Bangor and Ballyholme Bays, the submerged peat has been recorded on the beaches between low and high watermark (Praeger, 1892, p. 232; M'Skimin, 1811, p. 111). At Ballyholme, the peat is 4 to 6 inches (10 to 15 cm) thick and rests on a loose shingly breccia derived from the boulder clay (McHenry *in* Lamplugh and others, 1904, p. 110).

Submerged peat with some tree stumps in their position of growth, was also observed during the present resurvey on the coast about 1.6 km S of Donaghadee on Ballyvester Strand [594 781] where about 30 cm of peat is occasionally exposed between high and low water mark. On the published one-inch sheet this has been mapped as 'Submerged Forest'.

In the centre of Belfast Lough, when the Victoria Channel was being cut, Praeger (1892, pp. 234–235) noted about 30 ft (9 m) of *Thracia* Zone Estuarine Clay. The uppermost 13 ft (4 m) was stated to be more littoral in character than that from greater depths. What may be *Scrobicularia* Zone clay was encountered 35 ft (10.7 m) below high water mark. *Turritella communis*, *Abra alba*, *Mysia undata*, *Nucula nucleus*, and *Parvicardium exiguum* were recorded from this level [nomenclature revised].

On the north side of the Lough, an outlying deposit of Estuarine Clay has been described by McMillan (1947, pp. 16–19). The clay outcrops on the shore about L.W.M. north of Greenisland [385 852] and is visible along the shore towards Belfast for a distance of several hundred metres.

Section at Greenisland House

[Recent]	Old *Zostera* Bed	5 in (0.13 m)
[Upper Estuarine Clay]	*Thracia* Clay	13 in (0.33 m)
		(not penetrated)

McMillan obtained 58 species of mollusca from the clays of which five had not previously been noted. They were *Chrysallida spiralis*, *Clathrus clathratulus*, *Hermania catena*, *Philbertea linearis* and *P. purpurea*.

At the seaward end of a low reef running out below 'Raven Hill' [383 851], the clay yielded in abundance *Pecten maximus*, *Chlamys varia* and *C. distorta*. At the extreme low water on the seaward side of Greenisland, McMillan noted a saucer-shaped hollow of probably Estuarine Clay.

Submerged peat underlying beach gravel formerly occurred at Carrickfergus (Praeger, 1892, p. 231) between high and low tides. Calcified hazel nuts have been obtained from this peat together with willow and elder timber.

A small patch of peat, underlying typical *Scrobicularia* clay and resting on boulder clay, formerly occurred between tides on the shore close to Kilroot Station (Praeger, 1892, p. 231) where it was first noted by Stewart (1871, p. 27). The peat contained hazel nuts and stumps of Scotch fir *in situ*. The estuarine clay yielded *Macoma balthica*, *Scrobicularia plana* and a few other littoral shells.

This section has recently been re-exposed by the collapse of a sea wall and Estuarine Clay is seen beneath about 2 m of raised beach sand and gravels. The clays are grey, silty, and contain layers of shells and plant debris. the fauna recovered is:

Gastropoda: *Hydrobia ulvae*, *Littorina littoralis*, *L. littorea*, *L. sp.*, *Rissoa rufilabrum*, *Turboella radiata*.
Bivalvia: *Anomia ephippium*, *Cerastoderma edule*, *Chlamys varia*, *Heteranomia squamula*, *Ostrea edulis*, *O. sp.*, *Macoma* cf. *balthica*, *Parvicardium exiguum*, *P. sp.*, *Scrobicularia plana*, *Venerupis pullastra*.

This is an intertidal to shallow-water fauna containing forms tolerant of lowered salinities, as in estuary conditions. Some specimens show attrition but most appear not to have been transported to any significant extent.

Palynological examination of a specimen of the clay showed arboreal pollen, dominated by the genera *Pinus* and *Corylus* followed by *Quercus* and *Ulmus?* and a limited assemblage of organic-walled microplankton (dinoflagellate cysts) including *Spiniferites* cf. *ramosus*, *S. sp. nov.* [Cat. No. 023], *S. sp.* [undet.] and *Peridinium* cf. *oblongum*. The mixture of dinoflagellates and pollen is expected in an estuarine environment.

Estuarine Clay is known to underlie Larne Lough but has not been recorded further south than Magheramorne. The small area of salt-marsh within the area of this sheet is mapped as estuarine alluvium, but its age is not known though it is probably very recent.

RAISED BEACH

From the western margin of the map to Carrickfergus, the lowest raised beach is present along the whole coast and extends inland from the present day storm-beach to a cliff of boulder clay. An excellent temporary section in raised beach deposits was exposed during the excavations for the Carrickfergus sewage works west of Thompson's Point [398 864]. This revealed a basal bed of large rounded basalt cobbles with occasional chalk boulders, 15 to 30 cm in diameter, on which rested about 1.5 m of unbedded buff-coloured calcareous sand. Within this sand, particularly towards the sea, were several bands of cobbles generally consisting of a single layer of cobbles. The beach deposits, for the most part, rest directly on a wave-cut platform of Trias but towards the back of the beach they may occasionally be seen to overlie boulder clay.

At Carrickfergus, sands and gravels of raised beach age have been encountered in excavations in the town. Some Mesolithic implements are known (Stephens and Collins, 1960, fig. 1).

From Carrickfergus eastwards, patches of raised beach are present, though without any distinctive landward marginal feature.

South of Bonnybefore, just east of Carrickfergus, recent erosion has removed much of the raised beach promontory, whilst the shelly gravels immediately east of the former International Salt Works have now been built over.

Hull (1876, p. 35) recorded the raised terrace near Kilroot Station as having been worked in gravel pits, and in coastal sections where it overlay grey (estuarine) clays. A section seen in 1973 at Kilroot, where the sea wall had fallen revealed:

m

Gravel, medium, well-rounded with much sand and clay matrix 1
Gravel, coarse, well-rounded boulders up to 0.5 m in
    unbedded sand and gravel matrix. Pebbles lie flat   2

Sand, brown	0.02
Sand, grey silty	0.02
Peat, laminated silty	0.02
Sand, grey	0.1
Mud, grey silty with thin organic bands. Shells and plant debris, including hazel nuts	2.0

Shells from the raised beach deposits have been listed by Hull (1876) and Praeger (1897).

During the original survey in 1867–68, Du Noyer (1868, p. 495; 1869b, pp. 48–50) discovered Mesolithic (Palaeolithic of Hull, 1876, p. 35) flint implements at the Kilroot Station site [435 884].

A site investigation borehole put down in 1960, 450 m E of Kilroot Station proved 2.1 m of sand and gravel resting directly on boulder clay. East of the mouth of the Kilroot River, a raised beach platform, apparently devoid of sand or gravel deposits, is cut into boulder clay and the frontal margin is covered by storm-beach gravels. However, at the portal of the Irish Salt Mining and Exploration Company's mine, sands, probably of raised beach age, were encountered beneath slumped boulder clay. Between this locality and Cloghan Point, erosion has left no trace of the raised beach for about 2 km, though the railway construction may have contributed to its obliteration. At Cloghan Point [467 902] a small remnant of raised beach gravel about 1.2 m thick was noted (Praeger, 1890).

Fragmentary traces of the raised beach level were mapped east of Whitehead and remnants of a raised beach rock platform occur along the lava cliffs 700 m to the north-north-east, whilst north of Cloghfin there is a remnant of raised beach, up to 100 m wide and 900 m long, though its landward margin is obscured by slumped boulder clay over much of its length. North of Black Cave [485 953], extensive landslipping has obliterated any traces of raised beaches.

Welch (1902, pp. 214–216) records that during the construction of the cliff walk (now in disrepair) caves were excavated and yielded mammalian bones (including Red Deer horns) birds and fish remains from calcreted deposits.

On the south side of Belfast Lough, two low escarpments back terraces at about 8 and 18 m OD. The lower of the two features is particularly well developed and occurs as an almost continuous low escarpment, except where broached by steep-sided river valleys or mantled by wind-blown sand. The upper feature is much less continuous and is only intermittently recognised to the west of Bangor, although some 'shoulders' on the seaward side of drumlins to the south of Orlock Point [560 838] may also be related to this sea level.

The lower or '25 ft Raised Beach' is mostly cut in boulder clay but in headlands such as Grey Point, Carnalea, Wilson's Point, Luke's Point and Orlock Point it is developed in Lower Palaeozoic rock and forms low rock cliffs while the upper feature is cut entirely in boulder clay.

Between Grey Point [458 834] and Sydenham, at the south-western limit of the sheet, a well developed low escarpment cut in sands, boulder clay and Lower Palaeozoic rocks indicates the position of the inner edge of the '25-ft Raised Beach'.

At the Kinnegar [385 783], Praeger (1897, pp. 34–35) showed that the sickle-shaped gravel bar running west for almost a kilometre from a slight promontory on the shore below Holywood, rested on sand and shells which in turn overlay thick Estuarine Clay. The section was:

Section at Kinnegar

Stratified Gravels	8–12 ft	2.4–3.6 m
Sand and shells	3 ft	1 m
Estuarine Clay		

Praeger recorded from the sand the following fauna [nomenclature revised]:

Bittium reticulatum, Buccinum undatum, Gibbula cineraria, Littorina littoralis, L. littorea, Nassarius reticulatus, Neptunea antiqua, Nucella lapillus, Ocenebra erinacea, Patella vulgata, Turritella communis, Anomia ephippium, Cerastoderma edule, Chamelea striatula, Dosinia exoleta, Ensis ensis, Macoma balthica, Mytilus edulis, Ostrea edulis, Spisula subtruncata, Venerupis aurea, V. decussata, V. pullastra and V. rhomboides.

At Helen's Bay [462 828], the feature marking the limit of the former shore-line is largely blanketed by blown sand, although well developed in the headlands to north and south. However, in the north-west corner of the bay, immediately south of Horse Rock and just above high water mark, 1.8 m of red stony clay was formerly exposed below storm-beach material. This clay was probably boulder clay in which the raised beach platform had been cut.

At Crawfordsburn [467 825] the low escarpment marking the rear of the raised beach swings inland into Crawfordsburn Glen and in the banks of the stream, 150 m S of the weir, boulder clay is exposed below alluvial gravels. As at Helen's Bay, the raised beach terrace is largely covered by shingle and sand with a mantle of recent blown sand.

Between Crawfordsburn and Ballyholme Bay, the escarpment marking the back of the raised beach is well developed and is cut in both boulder clay and rock. Where superficial material rests on the terrace, the deposits are mostly recent shingle, except near Carnalea Station [482 823] where a 10 to 15 cm thick shell bed, resting on the Lower Palaeozoic rocks some 1.8 m above high water mark (Praeger, 1897, pp. 33–34), may represent a remnant of raised beach material.

At the date of the resurvey in 1958, all that could be seen of the late and post-Glacial deposits in Ballyholme Bay, was a little sandy peat and occasional patches of red stony clay in isolated inliers protruding through the sandy foreshore, while the former sections in the raised beach deposits, recorded by previous workers, were obscured by the concrete promenade, Praeger (1897, p. 33) records that:

'Before the present sea-wall was built, the raised beach here overhung the strand as a cliff of sand and gravel twenty feet in height, inhabited by quantities of sand-martins. Shells are rare in this bed; but at one spot, in a sandy layer three feet below the surface, and fifteen feet above high water, I obtained *Ostrea edulis, Mactra subtruncata* [ie *Spisula subtruncata*], *Trochus cinerareus* [*Gibbula cineraria*], *Littorina obtusata* [*L. littoralis*]. The shells were in a very crumbling condition. The gravels, which lie in horizontal beds, rest, at about half-tide level, on a thin layer of blue clayey sand, representing probably the Estuarine Clay Zone. Below this is the well-known bed of submerged peat, only about six inches thick, but containing the upright stumps of Scotch fir and other trees, in their natural position. Below this is a thin layer of bluish sandy clay, very tough, and full of branches and roots, succeeded by fine red sand or fine red clay. To the westward the boulder-clay rises up from below this series.'

Lamplugh and others (1904, p. 110) noted that a: 'higher or storm-beach accumulation of sand (probably in part blown sand) with layers of well rounded, but flattish beach-shingle has been heaped up to, in some cases, the altitude of thirty-seven feet (11 m) above Ordnance Datum or at least twelve feet (3.5 m) above the topmost layers of the ordinary Raised Beach'. He also noted that the blue-grey clay below the peat rested on 'a thin layer of loose shingly breccia' derived from the underlying boulder clay and concluded that the peat had accumulated on a boulder clay surface.

Near Groomsport, raised beach deposits have been recorded below storm-beach and wind-blown sand in Cove Bay [541 834], immediately north of Groomsport House Caravan Park which is situated largely on the raised beach terrace.

Macdonald (1934, p. 91) measured a section exposed by winter erosion in the deposits backing the present-day beach as follows:

		Ft	in	(m)
1	Recent surface soil		16	(0.4)
2	Blown sand with land shells	3	7	(1.1)
3	Black layer, probably a Neolithic hearth		10	(0.25)
4	Blown sand coarser than layer (2) but with the same molluscan fauna		indeterminate	

Layer 3 contained *Buccinum undatum*, *Littorina littoralis*, *L. littorea*, *Patella vulgata* and *Venerupis rhomboides* together with bones of horse, ox, pig and goat.

Similar shell-bearing deposits resting between the raised beach escarpment and the present coastline have been recorded by both Praeger (1897) and Macdonald (1961) at Sandeel Bay [555 835] about 1 km E of Groomsport. These beds occur at the eastern end of an extensive terrace of wind-blown sand which stretches from Groomsport House in the west of Sandeel Bay and extends inland for up to 400 m, mantling the raised beach feature for a considerable distance.

From Orlock Head [360 838] southwards to The Warren [588 811] at Donaghadee, the raised beach escarpment parallels the present shore-line closely and no undoubted raised beach deposits were recorded resting on the wave-cut rock platform at about 6 to 8 m OD, although some probably do occur in gardens on the inland side of the coast road.

At The Warren, an extensive area of blown-sand and storm-beach material may, at least in part, represent raised beach deposits.

About 2 km S of Donaghadee, raised beach deposits occur in a well marked terrace at about 7 to 9 m OD backing Ballyvester Strand [593 777]. On the Strand, just below high water-mark, about 0.3 m of peat is occasionally exposed resting on red boulder clay. The escarpment behind the present-day beach is cut in about 9 m of gravel and sand of discoidal pebbles of Lower Palaeozoic grit with some chalk and basalt pebbles and occasional layers of coarse sand. These deposits, which were seen in a 5 m deep temporary section at Beach Lea, rest on and against a pronounced notch cut in boulder clay and were themselves terraced by the '25 ft Raised Beach' sea.

'65-ft' Raised beach

The upper feature is less continuous than the 25 ft Raised Beach feature but where best developed on Carnalea Golf Course [486 823] forms a low escarpment with its base at about 18 to 20 m OD.

Further west, in the grounds of Crawfordsburn House, a similar feature [472 818] running south-west from Sea Hill may represent the same sea erosion level, while in the vicinity of Rockport House [436 818] not only is there a pronounced break of slope at about the same level, but a patch of cobbles and boulders, possibly representing coarse storm-beach deposits, rests on the boulder clay in the grounds of Glencraig.

To the south-west of Holywood, a pronounced terrace occurs above and on the inland side of the coast road which is built on the 25-ft Raised Beach terrace. The former terrace rises gently from about 16.5 m at the seaward edge to around 18 m some 300 m inland and may be related either to the features seen further east at about this level or could be associated with different levels of Lake Belfast.

Recent submergence

Marine erosion now taking place at several localities indicates a slight submergence in recent times. Knowles (1914, pp. 90–91) pointed out that along the County Antrim side of Belfast Lough, the gravels of the early post-Glacial raised beach were being swept away by the sea. On the opposite side of the Lough, Staples (1869, p. 42) recorded similar observations with reference to the foreshore near Holywood. Probably the most startling evidence is cited by Andrews (1892; 1893) who describes how the stack of an old windmill pump in a disused sandstone quarry at Cultra Bay, 1.5 km NE of Holywood, was surrounded by 1 m of water and situated over 15 m from the shore at high tide. It was estimated that between 1829 and 1892 some 2 ha of land were worked away in this vicinity and that the sea encroached on the land by 30 to 45 m. In this context, however, it is relevant to point out that the original six-inch-to-one-mile map of the Ordnance Survey marks this 'stack' as a disused lighthouse and that the 1834 and 1896 editions of the map show no significant change in shore-line.

PEAT

On the high ground in County Antrim, patches of peat and peaty alluvium, usually small, occur in hollows both in the boulder clay and rocks. Somewhat larger areas of peat occur in the area to the north of Black Hill such as Rigg Moss, Lockstown Bog and Loughdu Bog, but no information about the thickness of the peat is available and at the time of the resurvey none was being worked.

In County Down, there are fairly extensive areas of flat and cultivated peat south-east of Six Road Ends.

Evidence of the former widespread occurrence of peat can be found in the existence of many peat banks which now serve as field boundaries near the top of some of the drumlins, as on Blaeberry Island [550 777]. In this particular case, some of the older inhabitants can even remember when peat was cut on the upper slopes of this hill some 8 m above the present level of the bog. As there is still a good 1.5 m of peat left in the bottom of the neighbouring hollow, even as recently as 1890 there must have been upwards of 9 m of peat in some places. In 'Life in Mid-17th Century Down', Harris records that peat was cut in these bogs and supplied most of the northern part of the county with fuel. There is little peat left and most commonly the alluvial clays and gravels underlying the peat have been ploughed into the last remaining few centimetres of peat and now support luxuriant pasture.   AEG

ALLUVIUM

In County Antrim, stretches of alluvium are to be found developed along the length of all the watercourses. They are most extensive where the streams are in boulder clay and are generally either very narrow or absent where the streams are cutting through the solid rock. Within the larger areas, slight terrace development has occurred but this is generally inconspicuous and tends to be complex. There is never more than one terrace seen at any one place but in passing upstream the alluvial flat tends to disappear and what was the first terrace becomes the flood plain. This process is well illustrated in the Raloo Water between 'Burnside' and the old flax mill.

The alluvium is composed of dark brown silty material and is probably derived from the surrounding boulder clay. There is frequently an admixture of vegetable matter.

In County Down, there are narrow alluvial belts along many of the streams, and much more extensive areas in the south of the area where peat has been cut off and the underlying flat alluvial clays drained and cultivated.

LANDSLIP

An extensive area of landslip occurs on the steep slope below the Knockagh escarpment. Most of the slipped material is Triassic mudstone but blocks of basalt of all sizes are found in the melange and show every sign of having slipped bodily downhill.

Plate 21   Coastal landslip topography, slipped masses of White Limestone and Lower Basalt have moved down over Liassic and upper Triassic mudstones. North of Black Cave, Island Magee   [NI 353]

The landslip area is up to 600 m wide and extends from a base level of about 90 m OD to a height of about 240 m at the base of the basalt cliffs. The Cretaceous rocks are not seen at outcrop, being covered by fallen basalt, but it is assumed that there is a considerable volume of groundwater issuing from the base of this formation as the landslip area is very wet.

The landslips are caused by the low strength of weathered Mercia Mudstone which, when wet, become liquid. It may be surmised that the slipped material is of no great thickness, possibly of the order of 20 m or so, but, lubricated by groundwater and percolating rainwater, this surface layer moves gradually downhill, carrying with it masses of the overlying beds. The process is normally slow, with limited rotational slips leaving a scalloped hillside and tilting, sometimes overturning, the basalt and chalk masses. After heavy rainfall, however, catastrophic mudflows sometimes burst out from the base of the landslip. One such, in September 1958, flowed forward for over 200 m across the grazing land north of Knock Lodge. This surge followed considerable slow movement in the area for some days, which had deflected a water pipe-line and a high-voltage electricity line. This affected an area of hillside 500 m long and 250 m wide with a vertical range of about 70 m. The mudflow broke out immediately west of the area of movement and at a level about half way up it, and was accompanied by a great volume of water. Slow movement in the area continued for some weeks but after the mudflow there was no major slipping. Clearly the outburst had relieved high pore-pressure in the water-logged mudstones and restored some measure of stability.

Examination of the area showed that only the main tension gashes at the top of the landslip area had the classical concave-downwards form. The other gashes were straight down-slope, or small convex-down features, the former showing tear movements of several tens of centimetres. The area is heavily wooded, and one of the earliest signs of movement was the snapping of tree roots.

A second area of landslip occurs below the scarp on the east side of Carneal Hill where there has been movement on an inlier of Triassic mudstones. These slips appear to be stable and no recent movement has been recorded.   HEW

CHAPTER 15

Geophysical investigations

Most of the geophysical surveys which have been effected within the area of the Carrickfergus (29) Sheet have been made in connection with regional investigations of Northern Ireland and its continental shelf. Detailed geophysical surveys have been confined to the vicinity of the Carrickfergus salt-field and to the site of the power-station at Kilroot.

The regional investigations have comprised magnetic and gravity surveys over land and sea, and shallow seismic surveys at sea. Detailed gravity investigations have been made of small areas around the halite subcrop north-east of Carrickfergus and at Red Hall near the northern boundary of the One-inch Sheet, while a few detailed magnetic surveys have been made in the same areas in attempts to locate Tertiary dykes which might affect working of the salt. Geophysical logging has been carried out in some of the boreholes drilled in the salt beds. Detailed gravity, magnetic, and resistivity surveys have also been used in the investigation of the site conditions at Kilroot power station. Both the intrusive and extrusive Tertiary igneous rocks of north-east Ireland have been the subject of much palaeomagnetic investigation of oriented rock-specimens over many years, and although it is not clear from the published accounts of this work whether any of the specimens were collected from the Carrickfergus district, the results are very relevant to interpretation of magnetic field surveys in the area.

The marine geophysical investigations of part of the area have been reported (Wright and others, 1971) and a similar report covering the remainder of the offshore area is in preparation. Only the gravity data obtained from the marine surveys will be discussed further in this account.

MAGNETIC SURVEYS

The first systematic measurements of various parameters of the natural magnetic field in Ireland appear to have been made in the first half of the 19th century by Lloyd, Sabine and Ross (1836), who carried out experiments at some thirty localities throughout the country. Much of this work has been repeated, and new stations established, at irregular intervals up to comparatively recent times by Rücker and Thorpe (1891, 1896), Walker (1919), and Murphy (1953). One of the new stations, which are of a semi-permanent nature, is in the Bangor area within the Carrickfergus One-inch Sheet.

Aeromagnetic survey

The aeromagnetic survey of the Carrickfergus Sheet was made during 1959 as part of a survey of the whole of Northern Ireland and adjacent sea areas, arranged and supervised by the Geological Survey of Great Britain (now Institute of Geological Sciences) for the Ministry of Commerce, Government of Northern Ireland. A Gulf magnetometer was used to make continuous measurements of total magnetic field during flights at a clearance of 1000 ft (305 m) above land or sea along north–south traverses spaced at 2 km interval and east–west tie lines at 10 km intervals. The results have been published as a map (Geological Survey of Northern Ireland, 1971).

Results obtained over the Carrickfergus Sheet are shown in Figure 27: the contour values represent total force magnetic anomalies in nanoteslas (nT) above a calculated linear regional field which implies increases by about 2.2 nT per kilometre northwards and about 0.3 nT per kilometre westwards. The anomalies over the basalts of the Antrim Lava Series are complex and strong, producing a total variation of about 800 nT on the Carrickfergus Sheet, due to the strong magnetisation of these rocks. Now there is very convincing evidence from palaeomagnetic studies of oriented rock samples (Hospers and Charlesworth, 1954; R. L. Wilson, 1959, 1961, 1970; and Morris, 1972), and from study of both ground and airborne magnetic anomaly configurations over Tertiary igneous rocks (Bullerwell, 1954, 1961a, 1966), that all of the extrusive and intrusive volcanic rocks of Tertiary age in the north of Ireland possess a strong remanent magnetisation which is roughly opposite in direction to that of the present geomagnetic field. This accounts for the predominantly negative anomalies over the basalts in the Carrickfergus district, since the Antrim basalts can be most simply represented magnetically as a horizontal sheet of material strongly magnetised in a reversed direction. The anomaly pattern over such a model should consist of a broad negative anomaly over the basalt outcrop, with steep gradients coinciding with its western, southern and eastern boundaries, and a narrow positive anomaly along its western, northern and eastern margins. However, this pattern is complicated in reality by faulting and non-magnetic interbasaltic horizons causing the basalts to behave magnetically as a series of sheets.

The lack of intense anomalies offshore from Black Cave, Black Head, and White Head indicates that the basalts at these localities do not extend out to sea for any appreciable distance.

Reversely magnetised Tertiary dykes would be expected to produce mainly negative aeromagnetic anomalies, of a linear nature, and it is somewhat surprising that neither the massive olivine-dolerite dyke at Carrickfergus Castle, nor that which follows the Craigantlet Fault for 2 km to the south of Belfast Lough, produce significant anomalies. Despite this it is thought that the NNW-trending elongate negative anomalies between Bangor and Donaghadee are produced by an unexposed Tertiary dyke or group of dykes. The maximum amplitude of these anomalies is some 65 nT and quantitative interpretation of a WSW–ENE profile through the −70 NT closure suggests that the anomalies are

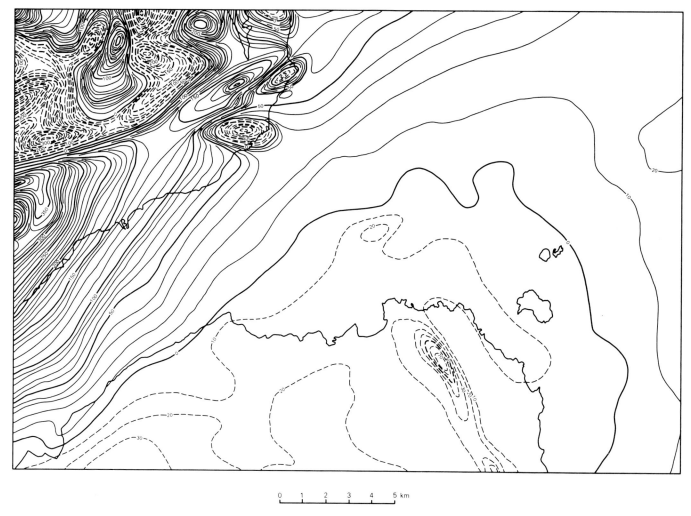

Figure 27 Map showing total force magnetic anomalies from aeromagnetic survey flown 1000 ft above terrain or sea

produced by a reversely magnetised, near vertical, sheet-like magnetic body with a NNW–SSE trend and great extent in depth. Precise estimation of the depth to the top of the body would not be reliable because the aeromagnetic flight paths crossed the anomalies very obliquely, but the depth is certainly below ground level. It should also be noted that these anomalies could be produced by a closely spaced group of parallel dykes whose anomalies would merge into each other at the flying height of the aeromagnetic survey.

The magnetic field over the rest of the Lower Palaeozoic strata of the mainland area south of Belfast Lough, the Copeland Islands, and North Channel is very smooth, indicating that these rocks are virtually non-magnetic and that the magnetic basement must lie at a depth of many kilometres in these areas. The strong magnetic gradient across Belfast Lough extends too far to the south-east to be associated with the Antrim Lava Group. It has been suggested that it may be related in some way to the continuation into Ireland of the Southern Uplands Fault of Scotland. Comments on the course of this fault in Ireland, based on geophysical evidence, have been frequent in recent

years (Murphy, 1955, p. 12; Bullerwell, 1961b, pp. 256–257; Leake, 1963, pp. 420–422; Anderson, 1965, pp. 388–389; Gunn, 1973, p. 111). Latterly the discussion has centred on the aeromagnetic evidence since it can be seen from the Aeromagnetic Map of Great Britain, Sheet 1 (Institute of Geological Sciences, 1972) that distinct magnetic anomalies occur along or near the Southern Uplands Fault throughout most of its length in Scotland. However, careful comparison of the location of the magnetic anomalies in relation to the mapped position of the fault shows that the anomalies are not produced by the fault itself, but by strongly magnetised igneous rocks to the north-west side of the fault. The igneous rocks which produce the magnetic anomalies appear to include the Ordovician Ballantrae Igneous Complex near the Scottish coast, as well as andesitic volcanic rocks of Old Red Sandstone age further to the north-east (Powell, 1970, p. 355). In the Ballantrae area the magnetic rocks at outcrop are at least 2 km from the fault.

The magnetic anomalies associated with the Ballantrae Igneous Complex appear to extend to the south-west across the North Channel, in the form of a magnetic 'high' with much reduced amplitude, to reach the Irish coast at Island

Magee. Powell (1970, p. 355) has estimated that burial of the complex under a 3 mile (4.8 km) cover of non-magnetic rocks in the North Channel would account for the fall-off in amplitude in that area. At Island Magee any onshore continuation of the anomalies is obliterated by complex anomalies produced by the Antrim Lava Series. However, the magnetic gradient across Belfast Lough appears to be the continuation of the gradient on the south-east side of the magnetic high across the North Channel. Moving north-west from Carrickfergus the increase in the magnetic field is masked by the strong negative anomalies at the edge of the Antrim Lava Group. Interpretation of the shape, form, magnetisation, and depth of the magnetic body producing the magnetic gradient across Belfast Lough is not possible because the whole anomaly is not seen, but one method described by Vacquier and others (1951), based on the horizontal extent of the maximum gradient over the edge of the body, suggests that in the area west of Carrickfergus the causative body may be of the order of 600 m below ground level. Also, the depth of burial appears to increase towards the north-east as the Mesozoic rocks thicken.

It is therefore suggested here that strongly polarised igneous rocks extend in a linear belt from the Ballantrae area across the North Channel into Ireland, where they are buried beneath Mesozoic and possibly Carboniferous sediments, and that they come relatively close to the surface in the Carrickfergus area. By analogy with the magnetic anomalies in Scotland the buried igneous rocks are probably an extension of the Ballantrae Igneous Complex but could even be of Old Red Sandstone age. Since the locations of both groups of igneous rocks in Scotland are not necessarily related to the Southern Uplands Fault, and since the spatial relationship between them and the fault is variable, the magnetic evidence cannot determine whether or not the Southern Uplands Fault extends into Ireland, or even indicate a possible course.

Ground magnetic surveys

Total field magnetic surveys have been carried out for Imperial Chemical Industries Ltd over small areas at Castle Dobbs and Red Hall. They successfully located the positions of igneous intrusions in both areas. Magnetic surveys were also used in the site investigation for the power station at Kilroot; these surveys were made both on land and from a boat offshore.

GRAVITY SURVEYS

The first measurements of gravity in the Carrickfergus area were made by Cooke and Murphy (1952) in 1950, when they established five gravity stations within the One-inch Sheet area, all of them south of Belfast Lough, during a regional survey of the north of Ireland. A more intensive survey of the land areas was made by geophysical staff of the Geological Survey of Great Britain (now part of the Institute of Geological Sciences) in 1959 in the course of a regional survey of the whole of Northern Ireland. An irregular grid of stations, averaging approximately one per square kilometre, was established in the area, using Wor-

den gravity meters. Five stations were established on Copeland Island, and one each on Light House and Mew Islands. The results of these surveys have been published by the Geological Survey of Northern Ireland (1967).

During 1969 and 1970, staff of the Marine Geophysics Unit of the Institute made gravity surveys of the sea area of the Carrickfergus Sheet, during regional geophysical surveys of the north Irish Sea and North Channel. A LaCoste and Romberg air-sea gravity meter, mounted in *mv* Moray Firth IV in 1969 and *mv* Surveyor in 1970, was used to obtain an analogue record of gravity along traverse lines. Position fixing was achieved using Main Chain Decca, and the accuracy of fixes varied from ±200 m to ±300 m. Unfortunately, it was not found practical to traverse along the length of Belfast Lough.

In 1967 N. D. Irwin established some 560 gravity stations in the Kilroot area as a project for a MSc degree of the Imperial College of Science and Technology. The field work was financed by the Geological Survey of Northern Ireland. The gravity measurements were made along roughly north–south traverses across the supposed positions of halite subcrops over an area of about 4 km². Similar gravity investigations have been carried out on behalf of Imperial Chemical Industries Ltd, in the vicinity of Castle Dobbs and Red Hall. The primary object of all these surveys was to detect mass deficiencies associated with halite beds. Pure halite has a density of only 2.0 g/cm³ whereas the average density of non-evaporite Triassic rocks is about 2.4 g/cm³. In fact, formation density logs from boreholes through the halite beds in the Carrickfergus area show that the density contrast between those beds and the mudstones in the Mercia Mudstone Group is of the order of -0.3 g/cm³. A vertical thickness of about 80 m of halite beds should therefore produce a negative gravity anomaly of 1 mGal. However, the halite beds in the Carrickfergus district occur in areas of relatively steep gravity gradients associated with thickening of the Permo-Triassic strata. The negative gravity anomalies produced by the Permo-Triassic basin are of the same sense as those to be expected from the halite beds, so it is therefore difficult to isolate residual anomalies produced by the latter. As a result the detailed gravity surveys have been generally unsuccessful in detecting the halite beds.

A short gravity traverse was established in the Kilroot area in 1973 by Wimpey Laboratories Ltd in connection with the investigation of the site of the proposed power station.

Bouguer anomaly map

The results of the 1959 land survey and the seaborne surveys are combined in Figure 28, as a map showing contours of Bouguer anomalies computed against the 1930 International Gravity Formula at sea level, referred to a gravity datum value 981.2650 cm/s² at Pendulum House, Cambridge. For land stations the necessary height reductions were based on density values estimated as appropriate to the geological strata between each station and mean sea level. The density values used were those accepted for use throughout Northern Ireland (for sources, *see* Bullerwell, 1961a, p. 238):

	g/cm^3
Antrim Lava Group	2.85
Ulster White Limestone	2.62
Permian and Trias	2.40
Carboniferous	2.65
Ordovician and Silurian	2.70

For sea stations a standard rock density of 2.67 g/cm³ was used for the Bouguer correction.

Although the seaborne gravity meter used produces a continuous record of gravity variation, reductions to the Bouguer anomaly were performed only at time intervals of thirty minutes for the 1969 survey, and ten minutes for the later survey. The positions of these reductions are shown on the Bouguer anomaly map. The accuracy of the anomaly values at sea, relative to local land gravity base stations, appears to be about ±2 mGal (Wright and others, 1971, p. 3), so they are contoured at 5 mGal intervals only in the sea area.

The main features of the Bouguer anomaly map consist of a broad area of relatively undisturbed high gravity field, with local maxima rising to more than ±27 mGal, over the Lower Palaeozoic rocks of north Down, the Copeland Islands, and adjacent sea areas; with much lower gravity to the north-west in the area occupied by Mesozoic strata and the basalts of the Antrim Lava Series. The low gravity anomaly reflects the mass deficiency associated with a deep basin of relatively light Mesozoic and possibly Carboniferous sediments underlying the basalts. A borehole has been drilled at Larne, near the centre of this basin, 6.5 km N of the Carrickfergus Sheet. There the Bouguer anomaly is about +5 mGal and the borehole was still in Triassic strata when terminated at a depth of 1283.5 m (Manning and Wilson, 1975).

This gravity anomaly pattern is not analogous to that in adjacent areas of Scotland, where the background gravity field over the Midland Valley appears to be significantly higher than that over the Southern Uplands (McLean and Qureshi, 1966), albeit superimposed on a westerly rise of the Bouguer and isostatic anomalies throughout western Scotland (Bullard and Jolly, 1936). But in Northern Ireland the background gravity field in County Down, the equivalent of the Southern Uplands in Scotland, is high while low gravity anomalies produced by the deep Mesozoic basin occur to the north-west where an even higher background

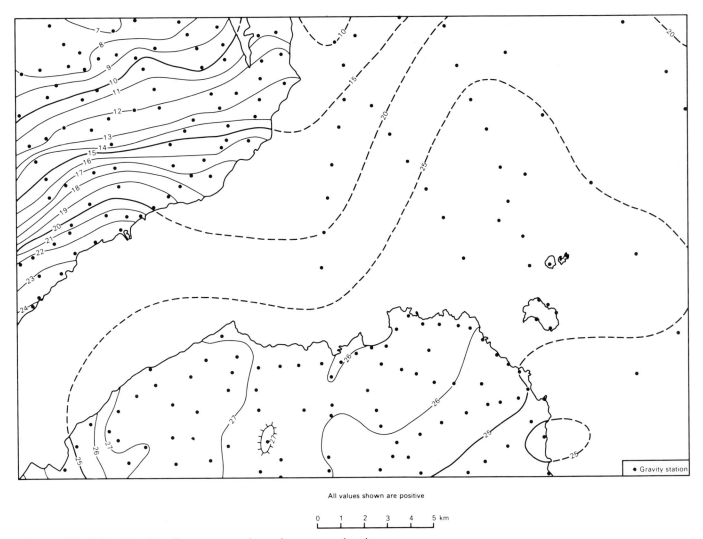

All values shown are positive

0 1 2 3 4 5 km

Figure 28   Map showing Bouguer gravity values at sea level

would be expected by analogy with Scotland. It is therefore difficult to determine the total gravity effect of the Mesozoic basin accurately. However, the background field north of Belfast Lough is unlikely to be lower than that to the south, and is more likely greater, so the observed gravity difference over the Mesozoic basin can be assumed to be a minimum value for the gravity effect of the light sediments above the Lower Palaeozoic basement. On this basis, the depth of the sedimentary basin in the Larne area must be at least about 2500 m, if the possibility of an anomaly due to Carboniferous sediments is disregarded. Since the density of the Carboniferous rocks in Ireland falls between those of the Permo-Triassic and Lower Palaeozoic rocks, then if Carboniferous sediments are present beneath the Mesozoic strata in the basin the minimum calculated depth to the Lower Palaeozoic basement is even greater than 2500 m.

The low gravity gradients across the Carboniferous and Permo-Triassic outcrops along the southern shore of Belfast Lough, and across the Lough to Carrickfergus, suggest that in this area the light sediments form only a relatively thin veneer on the Lower Palaeozoic strata, with the rapid deepening of the sedimentary basin occurring along a north-east to south-west line north of Carrickfergus. The observed gravity anomaly at Carrickfergus indicates that the depth to the top of the Lower Palaeozoic there is of the order of 650 to 850 m: this is comparable to the suggested depth of 600 m to the top of a magnetic body in the same area, deduced from the aeromagnetic evidence in the magnetic section of this chapter. There is no obvious gravity effect related to this magnetic body which is thought to occur within the Lower Palaeozoic rocks.

Bullerwell (1961b, p. 257) has pointed out that strong gravity gradients do not occur across the suggested lines of continuation of the Southern Uplands Fault into Ireland. If present, the fault would be expected to produce a linear gravity gradient, falling toward the north-west since it downthrows to that side in Scotland, but such an anomaly would again be of the same sense as that due to the light Mesozoic and Carboniferous sediments banked against the denser Lower Palaeozoic strata. There is therefore no positive gravity evidence to suggest that the Southern Uplands Fault is a major feature in this area: indeed the apparently sinuous nature of the gravity gradients in the vicinity of Belfast Lough suggests that they are unlikely to be controlled by faulting.

The high gravity field over the Lower Palaeozoic strata of North Down continues out to sea beyond the Copeland Islands before beginning to fall away towards the north-east into a gravity 'low' elongated along the North Channel, which is probably occupied by a sedimentary basin on the Scottish side.

CONCLUSIONS

Although both the gravity and magnetic fields in the Belfast Lough area display trends near to that which would be expected from a continuation of the Southern Uplands Fault, there is no positive geophysical evidence defining the position of the fault, or even to indicate that it does continue into Ireland as an important structure. On the other hand there is a strong indication from the magnetic evidence that the Ballantrae Igneous Complex extends across the North Channel beneath younger cover into the Carrickfergus area.

The gravity field is dominated by the negative residual anomaly produced by the Mesozoic/Carboniferous trough lying to the north-west of the Lower Palaeozoic massif of County Down. The gravity measurements at sea show that this massif must extend beyond the Copeland Islands for several kilometres to the north-east before being buried beneath lighter younger sediments.

JRPB

CHAPTER 16

Economic geology

AGGREGATE

Both the basalt lavas in County Antrim and the Lower Palaeozoic greywackes of County Down are widely used as sources of aggregate and roadstone, and although few quarries exist within this area there is extensive quarrying to the west and south.

Within the sheet limits, a very large quarry in Silurian greywackes is (1980) operational at Craigantlet [447 769] and smaller quarries are working at Holywood [408 784] and Ballykeel [415 775]. There are also a number of abandoned workings, which could be reactivated at any time.

In County Antrim, the only quarry active in recent years was on Muldersleigh Hill, near Black Head, though there are a number of small derelict workings on the high ground north of Carrickfergus.

The greywackes make a good aggregate provided that large quantities of the interbedded shales are not included in the crushed material. Greywackes tend to have a high 'polished stone value' (PSV) and good 'abrasion resistance' (AAV) which makes them useful as surface dressing for highways.

The basalt lavas require close control in working to ensure that vesicular and decomposed material from the top and bottom of lava flows is not incorporated. As individual lavas vary in grain size, density, composition and extent of weathering, basalt aggregate is variable in quality but with competent quarry control is suitable for most applications as roadstone or concrete aggregate.

Lightweight aggregate

Some shales and slates, when heated to near fusion, have the property of 'bloating', when the emission of gas causes the partly melted material to expand to several times its original size. If cooled at this point, particles of the materials are full of small gas cavities but have a vitrified surface and are thus non-absorbent and much lighter than the original shale. Concrete made from such aggregate is lighter and has better thermal insulating qualities than normal concrete.

The disused 'slate' quarries at Ballygrainey [528 788], south of Bangor, provide a material suitable for bloating and the old quarries have been examined as a potential source of shale. A bulk sample tested gave a good-quality product, clean, strong, fully vitrified and unsintered. The shale bloated directly without additives at an optimum kiln temperature of 1200°C ± 30°C. Lithologically, the 'slate' is a mudstone which splits into slate-like flags and was formerly used for roofing purposes.

BRICK CLAYS

Bricks were made from local Glacial clays in three pits in County Down. The partially obscured remains of one of these occurs on Grove Hill [520 817] on the north side of the Donaghadee Road in the townland of Ballyholme; another lay immediately to the south of the road and is now completely built over. The third pit was on the site of the carpet factory at Ballynoe [574 805], to the west of Donaghadee.

The brick clay in both pits is a bright red slightly sandy boulder clay and the stones must have been removed by hand picking. Fuller descriptions of these sections are given in Chapter 13.

Triassic mudstones were worked in an extensive pit [405 877] south of the railway near Clipperstown Halt, Carrickfergus, where a face of 6 m of mudstone beneath up to 4.5 m of boulder clay was exposed. These workings, in gypsiferous beds, must have produced a brick of only moderate quality, and working ceased before 1950. More recently mudstones were quarried from a pit [373 868] in the landslip area below the Knockagh escarpment and transported to a brick works in Belfast, but this ceased in the 1960's.

HYDROCARBONS

The possibility that deeply buried coal seams may exist beneath the Mesozoic rocks of the Larne basin and beneath bedded halites, implies that accumulations of natural gas could occur in the outcrop of the Mesozoic and Tertiary rocks. From 1967 to 1975 this area was held under a petroleum exploration licence by Marathon Petroleum (UK) Ltd and one borehole was drilled at Newmill, 4 km NNE of White Head [461 948]. Further details of this operation are not available for publication.

IRON ORE

The Lower Palaeozoic grits are commonly hematite-stained in the vicinity of faults. In at least one area, on the coast to the west of Bangor, this low grade ore is said to have been worked and smelted until the 16th century, using local charcoal. The record of this activity is commemorated in the name Smelt Mill Bay (Harris, 1744), but the activity must have been on a very small scale and it seems more likely that ore was imported, probably from the Furness area or Cumberland.

LEAD

In the townlands of Conlig and Whitespots, 6 km S of Bangor, there are the remains of formerly extensive 19th century lead mines [492 768]. Only a small part of the lode on which these mines were based lies within Sheet 29. However, as the whole area around the mines, lying in both Sheet 29 and Sheet 37, was mapped during the present survey this account deals not only with the mines on Sheet 29 but also those on Sheet 37, collectively known as the Newtownards and Conlig Lead Mines.

The main lode

The sites of all the known shafts (Figure 29) lie along a pronounced topographic feature trending N7°W. At the southern end and between the North Engine Shaft and the Conlig Shaft this feature is a pronounced east-facing escarpment with, in the intervening area, between the Gin Shaft and the North Engine Shaft, a pronounced gully. To the north of the Conlig Shaft, in the valley between Tower Hill Wood and Runestone Hill, a number of trial shafts or pits are still recognisable. The main, essentially north to south feature, is intersected by numerous smaller features trending south-east and, in fact, at its southern end the main feature dies out in one of these rocky gullies. These north-east trending features are probably aligned along dextral wrench faults. Some 320 m and 450 m respectively to the west of the main feature are two parallel features in one of which a series of small water-logged depressions may represent former trial pits. The 19th century lead mines extend for just over 1.6 km from north to south and were entered by eight shafts, a ladder-way and an adit. The deepest shaft, some 366 m, was the Bog (or Newtownards) Shaft.

At the time of the survey in 1958 the mines had not been worked for over half a century and were inaccessible, though two shafts were open—the South Engine and the Conlig Shafts. The South Engine Shaft was flooded up to about 30 m from the surface and water could be heard flowing through it. The Conlig Shaft was open to about 36.5 m, at this level a wooden obstruction or platform appeared to exist.

The sequence of mineralisation and the nature of the gangue can be reconstructed from material in the mine tips, particularly at the North Engine Shaft. The country rock of Lower Palaeozoic grits and shales in the vicinity of the lode is brecciated and cemented by anastomosing quartz veins. In cavities in this quartz-cemented breccia, baryte with galena, sphalerite, and chalcopyrite occurs. There is abundant pyromorphite on the dumps near the North Engine Shaft, first recorded by Manning (1960, p. 167). Blocks of basalt also occur in large numbers, some showing evidence of brecciation, with anastomosing veins of galena-bearing baryte, but never any quartz veining.

From this evidence Lamplugh and others (1904, p. 123) considered that 'the lodes must have been concentrated into their present position not earlier than the intrusion of the dykes, ie during Tertiary times'. Fowler (1959, p. 32) also regarded the mineralisation as Tertiary. However, Moorbath (1962, p. 330); dated galena from Conlig at 420 ± 70 Ma, which would indicate a Caledonian age, corresponding with the earlier opinion of Cole and Hallisay (1924, p. 73).

Both the postulated Tertiary and Caledonian ages appear improbable, not only on the basis of the observed relationship of the galena to the other gangue minerals, but as galena mineralisation at Castleward, 30 km to the south (Moorbath, op. cit.), has been dated as Armorican, while galena also occurs in the Visean limestones at Cultra (p. 97). Furthermore, Moorbath's data for the Conlig

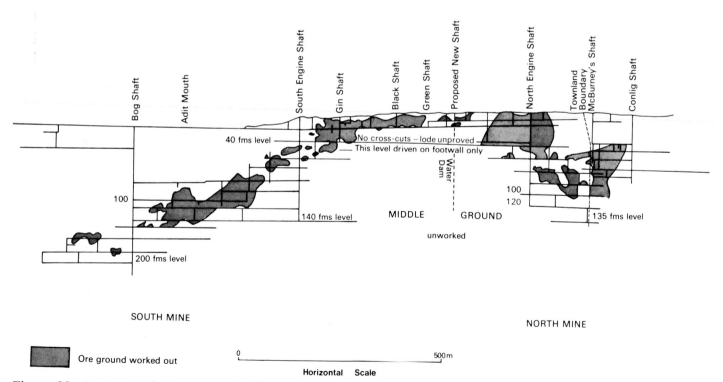

Figure 29 Section of the Newtownards–Conlig lead mines

galena can be recalculated (R. D. Beckinsale, personal communication) at 349Ma, using the Cumming and Richards (1975) model, or 266Ma, using the Stacey and Kramers (1975) model, and, in general, not much reliance can be placed on such model ages.

From observation of float material on the tip of the North Engine Shaft it is therefore postulated that the Conlig lead lode was emplaced in the following sequence:

1. Brecciation of the Silurian strata along the NNW-trending fault and quartz veining—Caledonian Orogeny.

2. Rebrecciation along the Caledonian fault and mineralisation with calcite, baryte and galena etc—Armorican Orogeny.

3. Intrusion of basalt dyke along pre-existing lode—Tertiary.

The following account of the lode is taken from the original Geological Survey memoir (Hull and others, 1871), based on the survey conducted in 1868 when, although the mines were inaccessible, the surveyors were able to gather information from men who had worked in the mines:

'The lode which has been worked under this name lies north of Newtownards, and has been proved to extend for the distance of over a mile in a general northerly direction; in its southern portion the lode runs N15°W for about 500 yds, and then trends more to the N, having throughout the remainder of its course a bearing of about N3°W. This mine was abandoned in 1865, so that at the time the district was under examination, it was impossible from the state of the workings to make a personal inspection of the lode; the following details, however, were gleaned from persons who had either been engaged in the working of the mine, or were on the spot when the mining operations were being carried out.'

From their account it would appear that the lode hades to the west at the rate of about one foot ten inches in the fathom, and coincides along its entire length with a dyke of dark green diorite [basalt], usually appearing along its walls, whilst the gangue of the lode is a fine angular breccia of Silurian rock, in a pale grey feldspathic matrix, through which there are strings of heavy spar, containing crystals of galena and minute quantities of copper pyrites, and peacock ore; some of these lead-bearing veins attaining a thickness of about ten feet.

Of the gangue of this mine, Dr Haughton, in a paper communicated to the Geological Society, Dublin, in 1852, thus writes:

'During a recent visit to Conlig lead mine, my attention was attracted by the peculiar appearance of the gangue of the lode, particularly by the asbestiform streaked appearance of the dark-green crystals forming the walls of the lode. This mine has been very rich in galena for some time past and was formerly very productive; but from about the 60-fathom level to the 90-fathom was comparatively unproductive. The gangue of the lode presents the same appearance from the surface to 120 fathom deep; it crumbles when exposed to the action of the air; is full of joint surfaces, which are coated with a mineral of the same chemical composition as the gangue itself, crystallised in an asbestiform manner, as shown by the accompanying specimens. The specific gravity of the gangue is 2.751; it fluxes readily before the blow-pipe into a black clay, and behaves in general as an ordinary trap rock.'

History of lead mining

A full account of the mines at Conlig and Whitespots is given by Woodrow (1978). The earliest record of lead mining in the area is contained in the 'Journal of the Progress of the mine works of the Bangor and Newtown Company'. In this journal James Millar notes:

'1780 October 6
Blue Lead Ore having been turned up by the plow in different parts of a field in Ballyleidy held of Sir John Blackwood by James Armstrong, broke Ground with five Miners, where Ore had appeared in greatest Bulk, in order to sink a Trial Shaft and found Lumps of Ore at Grass.'

Millar also records that after the Trial Shaft a second, the Comrade, was sunk as an air/water shaft 14.6 m to the east, the shafts being connected by a cross-cut at a depth of 16.5 m. No trace of these shafts has been found in Ballyleidy but in Ballyvarnet, some 292 m beyond the eastern limit of Ballyleidy and 228 m SE of Quebec Wood [482 788] two water-logged depressions, securely fenced, and now mostly filled with rubbish are referred to locally as mine shafts. These shafts, which are 14.6 m apart and on an east–west line, are probably the shafts referred to by Millar.

In the field to the east of the shafts and in the stone wall immediately to the north, field stones with traces of galena and iron pyrite embedded in barite are still found. The mineral vein, according to Millar's journal, hades to the north. Working on stringers of galena in the Ordovician grits, Millar records that some 24 tons of lead ore was produced over a period of one year at an expenditure of £368. There is no record of the ore having been smelted and it was probably shipped to Holywell in Wales where it would have fetched no more than £140. This small mine was abandoned on 29 September 1781.

A few weeks after commencing work on the Ballyleidy site Millar's journal also records the beginning of exploration further east:

'Newtown Works
There being in Mr Stewart's Deerpark, appearances of Trials made for the discovery of Ore and in some of them Ore having evidently been found; particularly at the place where four years ago a Course of eleven feet width, running N and S was discovered (by the Sinking of a Shaft 2 fath. deep and from its Bottom driving West under a Swamp to a distance of 17 fath.) out of which has been raised some Tons of Lead Ore Green and blue: the fall of the Ground favouring, it was judged advisable to drive a Level up to this work. A shoading (a system of trenches and pits cut through the overburden to the underlying solid rock) was accordingly commenced on the 22nd October with six Labourers accompanied by two Miners in order to trace the Vein. In the progress of the Shoading it was discovered that there are two main Courses, running N and S almost parallel to each other, inclined to intersect near the aforesaid Work and hading from the east. The 11 feet Course appearing to be the South one and blue Ore having been found in it, in the shoading, at 40 fath. distance from the Old Work, on entering the East it was determined to drive upon this Course.'

Having established the strike of the lodes Millar then opened a drift at the first place he had encountered the lode in the trenches and worked northwards. To the east of the drift he sank a series of air shafts to cut and prove the lode above drift level and finally met the drift by cross-cuts driven to the west. There must have been at least 6 air shafts on this level but no shafts belonging to this phase of working have been recognised. The entrance to the drift was probably somewhere in the area to the south of the

Windmill stump and the drift passed beneath the small lake [494 767] which was drained to enable working to take place below it. The precise life of this mine is not known as Millar's Journal stops in mid-sentence in June 1783. However, in a report to the Hibernian Mining Company in 1825 Thomas Weaver notes that this Journal was a partial transcript of an original document and that he had been informed that the mine was only worked for 7 or 8 years. In the 32 months covered by the Journal only 35½ tons of lead ore are recorded, 18 tons of which was raised in the last quarter.

In 1800, Donald Stewart, Itinerant Mineralogist to the Dublin Society, reporting on the mines, noted that 'a considerable number of tons of lead was raised by a company some years ago and a long level drove'.

The next record of mining is in an account book dated 1827–28 of the 'Newtownards Mines' operated by the Manx Cost-Book Company which has recently been discovered in the Isle of Man.

The earliest 19th century workings were on the northern portion of the lode and were carried on as two separate ventures; the North Engine House was erected in about 1837 to house a Cornish Beam Engine for a shaft ultimately going down to 120 fathoms (219 m). In the same year the Conlig Shaft [492 772] was sunk by the Ulster Mining Company under the direction of the famous mining engineer John Taylor of Norwich.

In 1842 George W. Dumbell of the Isle of Man took over the earlier 1827 mining operations, ie from the North Engine House southwards and developed the mines as the Newtownards Mining Company sinking the South Engine Shaft and the Bog Shaft in 1850 to depths of 256 and 366 m respectively.

Later in 1850 the Newtownards Mining Company bought out the Ulster Mining Company and operated both mines as the Newtownards and Ulster Mining Company, joining the mines below ground by a level.

Although the Londonderry Estate received royalties annually from 1829 until 1865 actual returns of ore are only available from 1845 (Mineral Statistics) and are as follows:

		Lead Metal	Silver
		ton	oz
1845	Newtownards	280	—
	Conlig	65	—
1846	Newtownards	137	—
	Conlig	42	—
1847	Newtownards	246	—
	Conlig	208	—
1848	Newtownards	366	—
	Conlig	179	—
1849	Newtownards	726	—
	Conlig	198	—
1850	Newtownards	671	—
	Conlig	189	—
1851	Newtownards	894	—
	Conlig	—	—
1852	Newtownards	1420	—
	Conlig	40	—
1853	Newtownards	1270	—
	Conlig	31	—
1854	Newtownards	1084	—
1855	,,	590	—

		Lead Metal	Silver
		ton	oz
1856	,,	430	—
1857	,,	363	—
1858	,,	242	485
1859	,,	170	340
1860	,,	128	—
1861	,,	—	—
1862	,,	240	481
1863	,,	208	416
1864	,,	97	194
1865	,,	23	46

Thus maximum production was reached in 1852 when 1460 tons of lead was obtained from both mines and after this production gradually declined until 1865.

Working on the royalty figures it is estimated that between 1828 and 1844 some 4000 tons of lead metal was produced, which added to the production recorded above indicate that some 13 500 tons of lead was mined over the period from 1828 until 1865. It is noteworthy that the Tamar Consols mine in Cornwall produced little more lead (14 640 tons) in 31 years and it was regarded as a major mine.

Between 1865 and 1880 no ore was produced but from 1880 until 1885 the Newtownards Mining Company raised small quantities of ore and a final attempt to reopen the mines was made in 1899 when 23 tons of lead, was produced (Cole, 1922, p. 94 records this as 82 tons).

Recent exploration of the area in the vicinity of the Newtownards and Conlig lead mines indicates that there is unlikely to be any economic orebody at reasonable depth.

AEG, AW

Other lead occurrences

Traces of lead mineralisation are also known from other localities within the area. In the deeply incised valley of the stream which separates the townlands of Carnalea and Ballykildaire small galena crystals occur on joints in the Ordovician rock about 10 m above high water mark [482 823].

In the 1871 Memoir this occurrence is recorded as 'about 100 yards N of Swinley Point where the stream which forms the western boundary of the townland of Carnalea enters the bay'. This reference is erroneous as Swinley Point juts out northwards into Belfast Lough some distance to the west of the stream referred to.

On the first edition of the field map no trace of lead is recorded at this locality but 'a thin seam of lead discovered forming conspicuous line' is noted trending north-eastwards on the western side of Smelt Mill Bay 200 m NW of the mouth of Bryans Burn.

Traces of galena were also found in quartz veins in Ordovician grits in the raised beach feature 200 m NW of Royal Belfast Golf Club House [429 817].

During the 1973 revision of the Carboniferous Ballycultra Formation (Division B) on the foreshore at Cultra galena was found to be a common mineral on joints in Bed 171, a compact micrite associated with the strike fault that repeats the outcrop of the beds west of the slipway and 210 m WSW of the Royal North of Ireland Yacht Club. The galena occurs as finely crystalline patches on joint surfaces coated in white calcite. A specimen of Bed 141, a calcareous sandstone from the same section, contained a discrete 2 mm crystal of galena. The presence of galena in this locality supports the view (expressed on p. 94) that the galena mineralisation is Armorican (or younger) rather than Caledonian.   AEG, AW

LIMESTONE

The Ulster White Limestone was formerly worked for lime-burning at many localities along the escarpment. All but one of these quarries have been abandoned. The currently active quarry, worked by the Antrim Grit and Lime Works Ltd near Redhall, produces lime and also grit from crushed flint. Just north of the one-inch Sheet a large cement industry is based on the Chalk at Magheramorne (Ballylig townland). The $CaCO_3$ content of the crushed chalk and flints, with some hand picking of the latter, is 94 per cent. A detailed analysis of this chalk is as follows:

Al_2O_3 (per cent)	0.15
Fe_2O_3	0.05
CaO	54.97
MgO	0.23
P_2O_5	0.16
H_2O+	0.63
H_2O-	0.10
CO_2	42.85
Soluble SiO_2	trace
Insoluble residue (mainly SiO_2)	0.92
Total	100.06

Analyst: M. H. Kerr

The Permian Magnesian Limestone at Cultra was worked at one time as a building stone—used in the architraves and corner stones of Carrickfergus Castle—and also, allegedly, as a source of dolomite for the manufacture of magnesium sulphate, for which it was exported to Glasgow. As the foreshore outcrop is limited this must have been on a small scale, and only by mining could any further dolomite be worked in this area.

PEAT

Formerly there were large areas of peat to the south of the road from Six-Road Ends to Donaghadee but there are now only small patches left. Turf from these bogs supplied fuel to large areas of north Down and as far back as 1744 Harris recorded that the Cottown Bog [555 785] was about 1000 acres in extent. From remnants of peat left on the upper slopes of drumlins the peat must have been at least 9 m thick in places. Many of the drumlins are still known as islands, eg Willy's Wood Island, and these names are descriptive of the time when only small hummocks of boulder clay stood up above the flat expanse of waterlogged peat.   HEW

SALT

Bedded halite (NaCl) in the Triassic Mercia Mudstone Group is known to underlie the area between Carrickfergus and Larne and has been worked for over 100 years. The mineral was first found in boreholes made, in search of coal, around the mid-19th century. The earliest reference to the discovery (Doyle, 1853, p. 232) refers to salt having been discovered near Glynn in 1839 in a borehole put down by a Mr Irving, 'when sinking (or boring) for coal'.

The subsequent discovery of rock-salt near Carrickfergus in a shaft being sunk in search of coal by the fourth Marquis of Downshire was probably in 1852 (Doyle, 1853, p. 232), though Miscampbell (1894, p. 546) claims that this discovery was made in 1845. Working of rock-salt commenced in the Duncrue area immediately after this discovery and later a group of mines was developed at Eden where a salt spring was known to occur. At Red Hall the mineral was only worked by solution mining (brining) from boreholes and no shafts were sunk.

Production of salt from brine continued until 1958 when the last works was abandoned. A new mine at Kilroot, 4 km ENE of Carrickfergus, worked from an inclined adit and producing rock-salt was started in 1967.

The maximum annual output from the earlier mines was 50 871 tons in 1912, and for many years production remained in the range 30 000 to 50 000 tons (Figure 30). After

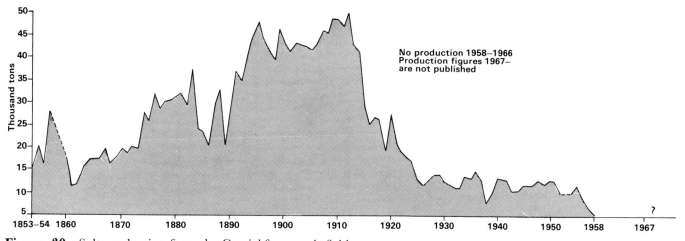

Figure 30   Salt production from the Carrickfergus salt field

1920 production declined from 10 000 to 15 000 tons and remained at this level until 1955. The substantial output figures from the Kilroot Mine are not published as this is the only mine producing rock-salt in Ireland. The history of mining in the four areas so far developed, ie Woodburn, Eden, Kilroot and Red Hall is as follows:

Woodburn area

There are some errors and ambiguities in early records of salt working in this area. Although rock-salt was discovered near Glynn in 1839 the mineral was not found at Carrick-fergus until 1852. An account of this discovery is given in a paper by Doyle (1853, pp. 232–236).

In the original Geological Survey Memoir of the area (Hull, 1876, p. 10) a section is erroneously recorded and claimed to be the succession at Duncrue. In fact, this is a composite section made up of the sequence in the Duncrue shaft, recorded by Kelly (1868, pp. 235–321), together with the lower portion of the succession drilled at Red Hall appended.

The section given by Doyle (1853, p. 234) for the Marquis of Downshire's shaft at Duncrue is as follows:

	Ft
Diluvium, about 50 ft;	
Red Marl, 500; intermixed with thin beds of gypsum	550
A thin stratum of rock-salt	15
Salt and Blue Band	6
Pure Salt	88
Blue and Red Band, with some salt	17
Mixed Salt, Blue and Red Band	13
Last Salt, clean but not yet bored through	20
Total depth	709

Another account of the discovery of salt at Carrickfergus is given in a much later account by Miscampbell (1894, p. 546) who claims that the original discovery was made in 1845, and he records that:

'A circular shaft was sunk, 9 ft in diameter, lined with brick to a depth of 750 ft, and from that point a boring was made a further 500 ft, but without finding coal. The labour, however was not fruitless for at a depth of 550 ft a workable seam of rock-salt was discovered, about 120 ft in thickness, with a couple of thin seams of mixed rock and marl interlaid.'

The actual date of the original discovery of rock-salt at Duncrue is further confused by the existence of an estate plan dated December 1848 (D671/M3/25) in the Downshire papers deposited in the Public Record Office of Northern Ireland. On this plan a salt works, designated the Duncrue Salt Works, is shown. However, it is probable that this note refers to a subsequent addition made to the map as amendments were made to the plan up until 1874.

That the original discovery was, in fact, made in 1852 would appear to be substantiated in the draft of a letter dated 16 November 1852 from the Marquis of Downshire to a group of Belfast businessmen, who were negotiating for rights to mine salt, in which he refers to the expenditure incurred in sinking 'within' this shaft. Irrespective of the date of the original discovery production of rock-salt from the Duncrue Mine began some time after the 9 March 1853

(after Doyle presented his paper) but before the end of the year, as production of about 15 000 tons from the 'neighbourhood of Belfast' in 1853–54 and of 20 000 tons from Duncrue in 1855 is recorded in Mineral Statistics for those years.

Although the title of the company which first worked the Duncrue Mine is not explicitly recorded, it is known that the group of businessmen who negotiated with the Marquis of Downshire on the 16 November 1853 were from Belfast and that by 1855 the mine was being worked by the Belfast Mining Company. It is thus probable that the Belfast Mining Company was established in 1852.

Following the meeting with the Marquis of Downshire but before the Duncrue Mine was successfully brought into production two abortive shafts were sunk by the mining company in search of salt closer to Carrickfergus and adjacent to the railway line to Belfast. The exact site of the first trial is not known but Miscampbell records (1894, p. 546):

'naturally the first consideration was to have their mine at the nearest possible point to a shipping port, and consequently a trial shaft was sunk about 1½ ml nearer the harbour of Carrickfergus (which is adjacent to the main line of railway) and due south from where the rock-salt had been found. Although the workings were carried to a much greater depth than the Company's advisers considered it necessary, the rock was not struck.'

The site of this trial is generally assumed to have been in the vicinity of Knockagh Cottage [390 874]. Miscampbell then goes on to record:

' a second trial shaft was then put down about a mile eastward of the first, with a similar result'.

The site of this shaft is shown on the 1857 edition of the Ordnance Survey 6-inch map [405 878], Antrim 52 and lies some 170 m NE of Woodburn Bridge at Clipperstown.

After these abortive attempts the mining company then returned to the vicinity of the Marquis of Downshire's original shaft at Duncrue and Miscampbell (1894, p. 546) records that:

'the Company were then reluctantly compelled to return to the neighbourhood of the original discovery, and at about 360 ft SW therefrom a couple of shafts were sunk.'

The distance quoted here, ie 360 ft (110 m) is quite inexplicable. The Duncrue shafts in fact lay 750 ft (229 m) SSE of Downshire's original shaft.

The Duncrue Mine produced salt from 1853 until it was abandoned in 1871 after it had become unsafe due to fracturing of the pillars supporting the roof. This crushing of the pillars was first observed in 1865 and as the crushing spread to adjacent supports it was decided to sink new shafts about 200 m to the north-north-east [393 893]. These shafts were completed in 1870 and the new mine, known as the French Park Mine, brought into production in the same year.

On the 26 April 1886 the Belfast Mining Company decided to go into voluntary liquidation and in 1887 their interest in the French Park Mine was purchased by Mr Alexander Miscampbell.

Miscampbell, according to McCrum (1909, p. 575), only

owned the French Park Mine for about one year before it, together with the Maiden Mount, Burleigh Hill, and Eden (Black Pit) mines were acquired by the Salt Union Ltd when he became manager of all four mines.

After the abandonment of the Duncrue Mine in 1871 the mine flooded and the area around the shafts caved in forming a roughly circular depression about 25 m across by 9 m deep.

By 1896 the workings of the French Park Mine were only 61 m from the flooded workings of the abandoned Duncrue Mine and a tunnel and pipe was driven through the separating wall so that brine could be abstracted from the flooded Duncrue workings. A second borehole for supplying freshwater to replace the brine abstracted in this way was completed on the 18 May 1898. Between 1896 and 1899 a considerable quantity of brine was abstracted from the Duncrue Mine via the horizontal boreholes and piped to a refining plant at Clipperstown. After about 9000 tons of salt had been removed in this way further subsidence of the surface took place over the Duncrue Mine culminating in an explosive outburst which hurled water, marl and debris some 18 to 20 m into the air and up to 100 m from the shaft (Rigby, 1905, p. 570). The site of the original subsidence over the Duncrue Mine is a wide hollow clearly defined on aerial photographs and on the Ordnance Survey map 6-inch sheet, Antrim 52. At present (1974) the depression [393 893] is being filled with urban garbage.

Withdrawal of brine from Duncrue continued even after the subsidence of 1899 and rock-salt production from the French Park Mine continued until 1937. In 1938 a collapse of the wall separating the Duncrue and French Park mines resulted in the latter flooding. Imperial Chemical Industries Ltd, who had only acquired the mine the previous year (1937) continued to abstract brine from the joined mines by pipes down the west shaft till 1958 when the workings were finally abandoned.

There were two other mines in this area, one at Maiden Mount [392 895] and the other at Burleigh Hill [398 898].

The Maiden Mount Mine was sunk by Mr M. R. Dalway in 1869 after he had carried out a series of unsuccessful sinkings at Eden [433 887] and in the vicinity of Glenview Cottage [400 894]. The first recorded production of salt from the Maiden Mount mine was not, however, until 1877. Some time prior to 1882, probably in 1880, Dalway also found salt in McKay's land in the north-east Division of Carrickfergus (McCrum, 1909, p. 515) which he worked for a time as the Burleigh Hill Mine before selling out to Messrs J. & W. Logan. The Logans carried out extensive improvements to the mine and erected a works at Bonnybefore for processing the rock-salt. Later this mine was sold back to Dalway who continued to operate both the Maiden Mount and the Burleigh Hill mines under the name of M. R. Dalway & Co Ltd until his interest was acquired in 1889 by the Salt Union Ltd.

Production of rock-salt from both the Burleigh Hill and Maiden Mount mines ceased in 1895 and Burleigh Hill was abandoned the following year. However, brine was abstracted from Maiden Mount until 1958.

The salt deposit worked in the Burleigh Hill Mine was apparently quite thin, some 7 m being worked and Miscampbell refers to the deposit as '. . . of small extent. The rock-salt is cut off on two sides of the mine, and altogether appears to be a mere pot'.

At Maiden Mount the salt which was worked was 39.2 m thick and occurred at a depth of 249 m.

Eden area

Although the first record of rock-salt having been produced in the Eden area, lying to the east of Carrickfergus, is not until 1884 (Mineral Statistics) when production of some 96 tons is recorded from 'Eden'. The existence of a salt spring had drawn attention to the potential of this area at a much earlier date.

In 1867 G. V. Du Noyer recorded the existence of an 'old' trial shaft [429 894] close to the salt spring marked on the 1857 edition of the 6-inch Ordnance Survey Sheet, Antrim 53. It was not, however, until 1866 that serious exploration began. In that year the Marquis of Downshire sank a shaft [433 887] 210 m NW of Eden Cottage. The shaft was dug to a depth of 162 m and an additional 100 m was bored but without finding rock-salt (or coal). Some 91 m below the surface a basalt dyke containing some brine was, however, encountered. The shaft was abandoned in 1867.

Mr M. R. Dalway also sank a borehole [428 894] in 1866 about 120 m WSW of the salt spring shown on the 1857 edition of the 6-inch Ordnance Survey map. Although this borehole was carried down to 107 m no rock-salt or brine was encountered but 30.5 m below the surface the marls contained some disseminated salt crystals.

The date of sinking the shafts to the 'Eden' mine, referred to in Mineral Statistics for 1884, is not known. This mine [429 894] lay some 130 m W of Trailcock Lane and 700 m NW of Eden School, close to the old shaft noted by Du Noyer. Access to the mine was by two shafts a few metres apart which have been variously referred to in subsequent accounts as the Eden No. 2, Black Pit or McAlister's Mine. Production from the mine was never large and in the first year (1884) only 96 tons of salt was recorded. The mine was apparently owned by a Mr Hodkinson before being acquired by the Salt Union Ltd in 1889 but was not worked thereafter and was finally abandoned in 1928. The more westerly shaft collapsed in July 1951 leaving a hollow some 12 m in diameter by 3.6 m deep and the remaining shaft partially collapsed a few months later, in December. Unfortunately, no plans of the extent of this mine are known to exist.

In 1890 another salt mine was opened some 400 m NNW of Eden School by the Chemical Salt Co Ltd. The main shareholder in this Company was Sir Charles Tennant of Peebleshire, who owned large chemical works in England and Scotland. This mine, the Tennant Mine [432 893], was operated by the Chemical Salt Company until it ceased production in 1921. Later, on the 11 February 1924 it was acquired by the Tharsis Sulphur and Copper Co Ltd who re-started work on the 18 February 1924. The final date of closure is not known.

The mine is still open having been used for mushroom production by the Monlough Mushroom Company at one stage. In 1965 it was acquired by the Irish Salt Mining and Exploration Co Ltd and although the shaft and winding gear have been refurbished no further working of salt has taken place.

The Tennant Mine was reached by a single shaft 103 m deep divided into two compartments so that two cages could operate. The floor dimensions of each cage were 1.4 m × 1.4 m. In the abandoned cavern some 7.6 m of salt were left forming the roof and the cavern is approximately 14.5 m in height. The pillars left to support the roof around the shaft bottom were marginally inadequate and the long term stability of the shaft is doubtful.

At about the same time that the Tennant Mine was being developed local men—Messrs William Vint & Sons— were developing another mine [431 896] some 250 m further to the north. This mine first appears in Mineral Statistics under the name of 'New Eden' in 1891. In 1892 the Salt Mines Syndicate was formed in Edinburgh by Mr Dundas Simpson who purchased the 'New Eden' mine from Messrs Vint. The mine continued to be worked by the Salt Mines Syndicate until March 1903 when it was taken over by Mr James Hodkinson, probably the same Mr Hodkinson who had owned the neighbouring Black Pit up until 1889. Mr Hodkinson died in July 1903 leaving the business to his four eldest sons who continued to manage the mine until about 1909 when a new type of salt refining process was introduced and the company was re-named the International Salt Company Ltd.

A few hundred metres to the west of the International Salt Company holding, a second mine [429 895] was opened in about 1894 and operated by the Carrickfergus Salt Works Company Ltd. A variety of names have been applied to this mine—it is sometimes referred to as the Downshire Mine but was also known as the Eden No. 1 or simply as the Carrickfergus Salt Company's Mine.

Work continued in both the Downshire Mine and the International Mines until only a thin wall of salt separated the two undertakings so that on the 9 July 1916 an unknown author reports 'on measuring this with a scraper the thickness of the barrier was only 9 inches'. Subsequently, after a series of contentious meetings, and possibly legal proceedings, the wall between the mines was removed and the operation carried on as one mine up until 1929. In 1929 the Salt Union Ltd obtained, from the Marquis of Downshire, 'rights of entry for searching and getting' salt from the joined mines but apparently neither the Salt Union nor Imperial Chemical Industries, who took over the Salt Union in 1937, ever worked the halite although the mines were kept on a care and maintenance basis until final abandonment in October 1958.

In the International Salt Company Mine two beds of salt were worked. The upper salt was 9.1 m thick and the lower salt 19.8 m. The thickness of the intervening strata is not recorded but is probably only, at most, one metre. Access to the mine was by a single shaft 2.4 m × 1.4 m divided to accommodate two similar cages, the descending cage partly balancing the rising one. Cole (1922) gave the shaft depth as 168 m but Geological Survey records show that the depth was 146 m.

The Downshire Mine was reached by two separate shafts each 1.2 m square and 6 m apart and were 152.4 m deep. The top of the worked salt lay some 131 m below the surface. In working this mine some 15 m of salt was left to form the roof of the cavern and only the lower 14 m was worked.

At various times during and after the Second World War the possibility of using both mines for storing explosives and foodstuffs was investigated but never pursued. In 1952 the mines were accessible and were inspected by two Civil Engineers, Messrs Huss and McGuigan of the Chief Engineer's Branch, Ministry of Finance. They recorded many useful facts about the condition of the mines and pointed out that at that time brine was flowing into the mines from behind a wooden stanking at a rate of some 10 000 gallons (455 000 litres) per week, which was removed by bailing. As ICI ceased to maintain the mines in October 1958 they are probably now filled with water, but it is not possible to verify their condition.   AEG

Kilroot area

No old works existed in this area but in 1967 the Irish Salt Mining and Exploration Co Ltd made two boreholes in the area south of Kilroot Church, one of which proved bedded halite near the surface. This area has since been developed by an extensive mine worked from an inclined adit and large quantities of crushed salt are exported from a 396 m long deep-water jetty. As this is the only producer of rock-salt in Northern Ireland production figures are not published.

Red Hall area

About 1500 m NNW of Ballycarry Station a brine pumping station [455 956] was marked on the 6-inch Ordnance Survey map near Oldmill House. Journals and a site plan obtained from the Geological Survey of Ireland show seven boreholes in and around this locality, some in Aldfreck townland and others in Redhall townland, with a further borehole 640 m SSE of the pumping station and one 900 m NNW in Ballyedward (one-inch Geological Sheet 21). The bores were sunk between 1892 and 1903, with a final diamond-cored hole in 1914. With the exception of the last, the holes penetrated only an upper series of halites some 100 to 140 m above the main salt beds.

Doyle (1853, p. 235) says that a new salt mine was discovered in Redhall about 1853 though Cole (1922, p. 133) observes that 'it may be noted that a "mine" in early usage often means merely a mass of mineral'. Curiously, Cole makes no mention of the above-mentioned boreholes (1892–1914) at Redhall, and it may be that those he recorded under the heading Magheramorne (Ballylig) have been confused with those now attributed to Redhall. The postal address of the operating company was Magheramorne.

There is little documentary evidence of the working of salt in this area, but it appears that the Larne Salt and Alkali Company carried on a brining operation, presumably after 1892. The works were on the seaward side of the railway at Drumnadregh, 2 km N of the brine pumping station and though derelict in 1917 were operating vacuum pans in 1924, as claimed in a letter to the Commission on the Natural and Industrial Resources of Northern Ireland. When they finally closed is not known.

Composition and quality of the salt

Few analyses of the salt are available from the Carrickfergus area, but so far as is known the soluble mineral in the evaporite beds is nearly all halite, (NaCl).

From the Larne analyses it will be seen that even with the elimination of insoluble residues, no more than about 95 per cent NaCl can be reached, the balance being sulphates and water, and run-of-the-mine rock-salt is likely to be in the range 91 to 93 per cent.

As in the Cheshire saltfield, the potassium present is largely in the form of clay minerals and not as soluble salts.

Table 6   Analysis of Antrim salt

J Chemico-Agricultural Soc., Ulster 1880, 14, 80		Miscampbell (1894, p. 566)		Composite samples Larne Borehole	
NaCl	91.83	NaCl	96.17	NaCl	89.9–75.7
CaSO$_4$	4.8	MgCl$_2$	0.03	Ca	0.7–0.16
MgCl$_2$	0.68	KCl	0.15	Mg	0.28–0.02
Clay	0.37	Al$_2$O$_3$	0.35	K	0.09–0.04
H$_2$O	2.32	SiO$_2$	0.82	SO$_4$	2.68–0.31
		H$_2$O	2.48	Insoluble	18.81–5.32
				H$_2$O	2.42–0.39

Reserves

Exploration has been concentrated in the four worked areas and there has been no attempt to evaluate the total resources of the field. Early writers such as Miscampbell (1894) considered that the Duncrue and Eden areas were discrete local saltfields and it is possible that the beds are, in fact, not continuous but are restricted to structural basins, although the halite-bearing groups increase in thickness northwards reaching c. 700 m at Larne. The largest reserves may therefore be expected in the northern part of the area.

The sub-drift outcrop of saliferous beds at Eden shown on the published one-inch geological map is now considered to be incorrect and the current concept of the distribution of the salt outcrop is discussed in Chapter 6.

Future mining development depends on continued demand for crushed rock salt and the ability of producers to meet increasingly stringent specifications for purity. Recent experimental work by the Department of Industrial Science, Lisburn suggests that calcining crushed halite with sodium silicate produces a less corrosive brine and converts marl impurities into a hard skid-resistant grit, and such treatment might increase the acceptability of moderate grade material.

The revival of the brined salt industry—solution mining—depends on the demands of industry for a brine feedstock, though the development of cavities for the storage of oil products and compressed air for peak-load power generation has been actively considered.

SAND AND GRAVEL

Sand and gravel occur at a number of places in the south-eastern part of the area and also west of Holywood.

Even the Hogstown 'gravel', which occurs in a moraine, is ill-sorted and was used only as ballast by the Belfast and County Down Railway. The Eagle Hill gravel, well bedded and obviously water lain, was used towards the end of the 1939–45 war when large quantitites were used in the construction of the neighbouring uncompleted military airfield.

The extensive deposits between Holywood and Mertoun Hall have been worked from a pit on the southern edge of the map, most recently as a source of fill. The material here is a fine red silty sand, too fine for most building applications and was used as a moulding sand.

WATER SUPPLY

Though used in one or two localities, groundwater is not plentiful in this district and most of the water for public water supply is obtained from surface catchments. Most of the streams flowing southwards towards Belfast Lough are impounded directly, or via aqueducts and tunnels, in a series of reservoirs at Woodburn, Lough Mourne and the Copeland River, with a total capacity of 1183 million gallons. Studies are being carried out of the effect of afforestation on these catchments which total about 2800 ha in extent, but no conclusions have been reached. Some of this water is piped to the Belfast area but the town of Carrickfergus and the major industrial plants in the vicinity are supplied from these sources.

Public supplies in County Down are from reservoirs at Portavo, Ballysallagh (2), Creighton's Green and Ballykeel (2). These reservoirs are situated in shallow valleys variably covered by drift resting on impervious Lower Palaeozoic rocks. Supplies from them are augmented by water from the Mourne aqueduct and from Newtownards (Sheet 37) where it is obtained from wells in Triassic sandstone and river gravels.

Groundwater

Lower Palaeozoic

The permeability of the slates, shales, mudstones and greywackes, is low, and though joints in the uppermost weathered zone may yield up to 700 gph, 60 to 80 gph is more common (Manning, 1971, p. 339). Most of the wells are dug wells and appear to derive their supplies from the stony base of the boulder clay and not from the solid rock into which there is commonly little penetration. The Lower Palaeozoic rocks must be regarded as a poor aquifer.

Carboniferous

No details of the potential yield of the two Carboniferous formations—the Craigavad Sandstone and the calcareous and argillaceous Ballycultra Formation—are known. Both formations have a largely sub-drift outcrop but the sandstones should be capable of moderate yields.

Permian

Though the Permian Upper Marls have no potential, the Magnesian Limestone and the underlying sandstones in the south-west of the district should have moderate yields, though the water will be very hard. A bore at Kinnegar in 1971 gave a yield of 6000 gph, from the Magnesian Limestone which was about 16 m thick. The water has 384 mg/litre of dissolved solids, 79.7 mg/l sulphate, and 1.46 mg/l fluorine.

Triassic

The Sherwood Sandstone Group constitutes a major aquifer in the Belfast area. Its outcrop within the sheet is, however, limited to two narrow coastal strips on the north and south shores of Belfast Lough. Between these strips, the floor of the Lough is largely underlain by sandstone.

On the north shore of the Lough, exploitation of the groundwater potential, with the possibility of salt-water intrusion, must be regarded as doubtful. To the south, however, the prospects are slightly better, though a well on the intake area near the oil refinery gave saline water. Sites near the edge of the outcrop might give moderate yields, though the possibility of saline contamination is always present.

The dull red mudstones with rare thin sandstones of the Mercia Mudstone Group are generally of low permeability and are not regarded as water-bearing, save where intruded by Tertiary dykes or vents which may act as conduits.

Cretaceous

The Ulster White Limestone (Chalk) is a hard non-porous rock in Northern Ireland and its value as an aquifer is limited to the water flowing through joints and fissures. It has a narrow outcrop within the area of the one-inch Sheet and the replenishment of groundwater must come from waters percolating through the overlying basalt. The basal Hibernian Greensands should constitute a good source of groundwater, and springs along the outcrop issue from the base of the greensands which rest on impermeable Lias or Triassic mudstones. From the few wells in the Cretaceous it

is not possible to estimate the relative contributions of the limestone and greensand, but data from elsewhere suggests that groundwater in this group of rocks is of recent origin and flows largely in open joints.

Springs at Sullatober, north of Carrickfergus, which supply water for the public system issue from the permeable Hibernian Greensand, and part of Whitehead's water supply is derived from wells sunk into the Cretaceous chalk and Hibernian Greensand.

Tertiary lavas and intrusives

Though individual flows of the Antrim Lava Series are well jointed, the inter-lava boles are relatively impermeable and most wells give poor yields. Nevertheless some recent wells sunk into the lavas suggest that there is a local groundwater potential. At Mossley (Sheet 28) up to 9000 gph were produced from a 6-in diameter 122 m deep borehole, and where the lavas are broken by faulting on low ground prospects are reasonable. A volcanic vent penetrated by an exploration borehole near Kilroot Church, is reputed to contain abundant fresh water.

Drift

The boulder clay varies considerably with the underlying rock, but is usually relatively impermeable. In the past many rural dwellings have utilised shallow wells in this deposit, the wells usually being sunk to rock-head. The large diameter (1.2 to 1.8 m) of these wells permits them to act as storage reservoirs even though the rate of percolation may be slow. Fortuitous sand and gravel lenses may produce a higher yield.

Though there are no documented examples from within the one-inch Sheet, alluvial flats may give large supplies. From a shallow excavation in an alluvial area near Irish Hill, for example, 4000 gph was pumped. PIM, HEW

REFERENCES

ADAMSON, J. H. and WILSON, G. F. 1933. Petrography of the Lower Carboniferous Rocks of N.E. Ireland. *Proc. R. Ir. Acad.*, Vol. 41B, pp. 179–190.

AGASSIZ, L. 1842. On glaciers and the evidence of their having once existed in Scotland, Ireland and England. *Proc. Geol. Soc. London*, Vol. 3, pp. 327–332.

ANDERSON, E. M. 1951. *The dynamics of faulting and dyke formation with applications to Britain.* 206 pp. (Edinburgh: Oliver and Boyd.)

ANDERSON, F. W. 1950. Some reef-building calcareous algae from the Carboniferous rocks of northern England and southern Scotland. *Proc. Yorkshire Geol. Soc.*, Vol. 28, pp. 5–28.

— and DUNHAM, K. C. 1966. The Geology of Northern Skye. *Mem. Geol. Surv. G.B.*

ANDERSON, J. 1873. On the Geological Formations of County Down. *Proc. Belfast Nat. Hist. Philos. Soc.*, 1871–72, pp. 41–49.

ANDERSON, T. B. 1964. *The stratigraphy, sedimentology and structure of the Silurian rocks of the Ards Peninsula, Co. Down.* University of Liverpool Ph.D. thesis (unpublished).

ANDERSON, T. B. 1965. The evidence for the Southern Uplands Fault in north-east Ireland. *Geol. Mag.*, Vol. 102, pp. 383–392.

ANDREWS, J. H. 1965. Christopher Saxton and Belfast Lough. *Ir. Geogr.*, Vol. 5, pp. 1–6.

ANDREWS, M. K. 1892. Denudation at Cultra, County Down. *Annu. Rep. Proc. Belfast Nat. Field Club*, Series (2), Vol. 6, pp. 529–532.

— 1893. Denudation at Cultra, County Down. *Ir. Nat.*, Vol. 2, pp. 16–18, 47–49.

ANON. 1931. Belfast and District Water Supply. *British Waterworks Assoc.*

ARCHER, F. 1881. Worked flints of the raised beaches of north-east Ireland. *Proc. Liverpool Geol. Soc.*, Vol. 4, pp. 209–216.

BADEN-POWELL, D. F. W. 1937. On the Holocene marine fauna from the implementiferous deposits of Island Magee, Co. Antrim. *J. Animal Ecol.*, Vol. 6, pp. 87–91, 274–276.

BAILEY, E. B. 1924. The desert shores of the Chalk Seas. *Geol. Mag.*, Vol. 61, pp. 102–116.

— 1924. *In* CLOUGH, C. T., HINXMAN, B. A. and OTHERS 1925. The geology of the Glasgow District. *Mem. Geol. Surv. Scot.*

— 1930. *In* RICHEY, J. E., ANDERSON, E. M. and MACGREGOR, A. G. 1930. The geology of North Ayrshire. *Mem. Geol. Surv. Scot.*

— 1874. Sketch of the geology of Belfast and the Neighbourhood. *Hardwick's Science Gossip*, pp. 169–170.

BARROIS C. 1876. Recherches sur le terrain Crétacé supérieur de l'Angleterre et de l'Irlande. *Mem. Soc. Geol. du Nord.*, Vol. 1.

BELFAST NATURALISTS' FIELD CLUB. 1888. Report of the committee appointed to investigate the Larne Gravels, etc. *Rep. Proc. Belfast Nat. Field Club*, Series 2, Vol. 2, (1886–7).

BELL, A. 1891. Fourth and final report of the committee appointed for the purpose of reporting on manure gravels of Wexford. *Proc. Belfast Nat. Field Club.*

BELT, E. S., FRESHNEY, E. C. and READ, W. A. 1967. Sedimentology of Carboniferous Cementstone Facies, British Isles and Eastern Canada. *J. Geol.*, Vol. 75, pp. 711–721.

BERGER, J. F. and CONYBEARE, W. 1816. On the geological features of the north-eastern counties of Ireland. *Trans. Geol. Soc. London* (1), Vol. 3, pp. 121–195.

BERRY, L. G. (Editor). 1970. Powder diffraction file. Set 20. Card 20–452. Joint Committee on Powder Diffraction Standards. Philadelphia, Pennsylvania.

BINNEY, E. W. 1855. On the Permian Beds of north-west England. *Mem. Litt. Philos. Soc. Manchester*, 2, Vol. 12, pp. 209–269.

BISHOPP, D. W. and OTHERS. 1948. The geology of Eastern Ireland. *Int. Geol. Congr. 18th Session. London.*

BLACK, M. 1953. The Constitution of the Chalk. *Proc. Geol. Soc. London*, No. 1499, (Session 1952–3), pp. 81–86.

BOSENCE, D. W. J. 1973 Recent serpulid reefs, Connemara, Eire. *Nature, London*, Vol. 242, pp. 40–41.

BLOXHAM, T. W. and ALLEN, J. B. 1959 Glaucophane-schist, eclogite and associated rock from Knockormal in the Girvan–Ballantrae complex, South Ayrshire. *Trans. R. Soc. Edinburgh*, Vol. 64, pp. 1–27.

BRADY, G. S., CROSSKEY, H. W. and ROBERTSON, D. 1874. Monograph of the Post-tertiary Entomostraca of Scotland, including species from England and Ireland. *Monogr. Palaeontol. Soc.* XXVIII.

BRANDON, A. 1977. The Meenymore Formation—an extensive, intertidal evaporitic formation in the Upper Visean (B₂) of north-west Ireland. *Rep. Inst. Geol. Sci.*, No. 77/23, 14 pp.

BRINDLEY, J. C. 1967. The geology of the Irish Sea area. *Ir. Nat. J.*, Vol. 15, pp. 245–249.

BROWN, W. O. Some soil formations of the basaltic region of north-east Ireland, *Ir. Nat. J.*, Vol. 11, pp. 120–132.

BRYCE, J. 1837a. On the geological structure of the north-eastern part of the County of Antrim. *Trans. Geol. Soc. London*, (2), Vol. 5, pp. 69–81.

— 1837b. On the Magnesian Limestone and associated beds which occur at Holywood in the County of Down. *J. Geol. Soc. Dublin*, Vol. 1, pp. 175–180.

— 1852. Geological Notices on the environs of Belfast, the East coast of Antrim, and the Giant's Causeway. Special pamphlet prepared for the Belfast meeting of the British Association in 1852.

— 1853. On the geological structure of the Counties of Down and Antrim. *Rep. Brit. Assoc.* [for 1852], Vol. 22, pp. 42–43.

BULLARD, E. C. and JOLLY, H. L. P. 1936. Gravity measurements in Great Britain. *Monogr. Notes R. Astron. Soc. Geophys. Suppl.*, Vol. 3, pp. 443–477.

BULLERWELL, W. 1954. A vertical force magnetic survey of the Coalisland district, Co. Tyrone, Northern Ireland. *Bull. Geol. Surv. G.B.*, No. 6, pp. 21–32.

— 1961a. *In* FOWLER, A. and ROBBIE, J. A. 1961. Geology of the country around Dungannon. *Mem. Geol. Surv. North. Irel.*

— 1961b. The gravity map of Northern Ireland. *Ir. Nat. J.*, Vol. 13, pp. 254–257.

— 1964. Geophysical and Drilling Exploration in Northern Ireland. Progress Report. *Govt. of North. Irel.* (Belfast: HMSO.)

— 1966. *In* WILSON, H. E. and ROBBIE, J. A. 1966. Geology of the country around Ballycastle. *Mem. Geol. Surv. North. Irel.*

BURCHELL, J. P. T. 1931. Early Neanthropic man and his relation to the Ice Age. *Proc. Prehist. Soc. East Anglia*, Vol. 6, pp. 253–263.

— 1932. Implements of Late Magdalenian age underlying the raised beach at Larne, County Antrim. *Nature, London*, Vol. 29, p. 726.

— 1933. Flint implements of Early Magdalenian age from deposits underlying the Lower Estuarine clay at Island Magee, Co. Antrim. *Nature, London*, Vol. 132, p. 860.

— 1934. Some littoral sites of early post-glacial times located in Northern Ireland. *Proc. Prehist. Soc. East Anglia*, Vol. 7, (3), pp. 366–372.

— and WHEELAN, C. F. 1930. Palaeolithic man in north-east Ireland. *Nature, London*, Vol. 126, p. 352.

CAMERON, T. D. J. 1977. Silurian metabentonites in Co. Down. *In* Palaeozoic volcanism in Great Britain and Ireland. Conference Report. *J. Geol. Soc. London*, Vol. 133, p. 404.

— and OLD, R. A. In prep. Geology of the country around Meenymore. *Mem. Geol. Surv. North. Irel.*

CARRUTHERS, R. G. 1927. The oil-shales of the Lothians. *Mem. Geol. Surv. Scot.*

CHARLESWORTH, J. K. 1936. Geomorphology of the Irish Sea Basin. *Nature, London*, Vol. 137, pp. 1040–1041.

— 1937. Recent Progress in Irish Geology. *Ir. Nat. J.*, Vol. 6, pp. 265–273.

— 1939. Some observations on the glaciation of north-east Ireland. *Proc. R. Ir. Acad.*, Vol. 45B, pp. 255–295.

— 1950. Recent Progress in Irish Geology. *Ir. Nat. J.*, Vol. 10, pp. 61–71.

— 1953. *The Geology of Ireland. An Introduction.* 276 pp. (Edinburgh and London: Oliver and Boyd.)

— 1957. *The Quaternary Era.* 1686 pp. (London: Arnold.)

— 1963a. Some observations on the Irish Pleistocene. *Proc. R. Ir. Acad.*, Vol. 62B, pp. 295–322.

— 1963b. *Historical geology of Ireland.* 565 pp. (Edinburgh and London: Oliver and Boyd.)

— 1973. Stages in the dissolution of the last ice-sheet in Ireland and the Irish Sea region. *Proc. R. Ir. Acad.*, Vol. 73B, pp. 79–86.

— and HARTLEY, J. J. 1935. The Tardree and Hillsborough dyke swarms. *Ir. Nat. J.*, Vol. 5, pp. 193–196.

— and PRESTON, J. 1960. *Geology around the University Towns; north-east Ireland; the Belfast Area. Geol. Assoc. Guide*, No. 18, 30 pp. (Colchester: Benham.)

— and OTHERS. 1935. The geology of north-east Ireland. *Proc. Geol. Assoc.*, Vol. 46, pp. 441–486.

— and OTHERS. 1960. The geology of north-east Ireland. *Proc. Geol. Assoc.*, Vol. 71, pp. 429–459.

CLAPHAM, M. 1928. The diatoms of Woodburn Glen. *Ir. Nat. J.*, Vol. 2, p. 93.

CLARK, R. 1902a. Notes on the fossils of the Silurian area of north-east Ireland. *Geol. Mag.*, (4), Vol. 9, pp. 497–500.

— 1902b. The Silurians of north-east Ireland and their characteristic fossils. *Ir. Nat.*, Vol. 11, p. 274 [Abstract B.A. paper].

CLELAND, A. McI. 1932. White basalt. *Ir. Nat. J.*, Vol. 4, pp. 93–94.

— 1933. An interesting rock exposure at Whitehead. *Ir. Nat. J.*, Vol. 4, pp. 209–211.

— 1938. Red flints. *Ir. Nat. J.*, Vol. 7, pp. 5–8.

CLOSE, M. H. 1866. Notes on the general glaciation of Ireland. *J. R. Geol. Soc. Irel.*, Vol. 1, pp. 207–242.

COFFEY, G. and PRAEGER, R. L. 1904. The Antrim raised beach. A contribution to the Neolithic history of the north of Ireland. *Proc. R. Ir. Acad.*, Vol. 25, pp. 143–200.

COLE, G. A. J. 1922. Memoir and map of localities and minerals of economic importance and metalliferous mines in Ireland. *Mem. Geol. Surv. Irel.*

— and HALLISSY, T. 1924. *Handbook of the Geology of Ireland.* 82 pp. (London: Murby.)

COOKE, A. H. 1950. Measurements of gravity in Ireland. Pendulum observations at Dublin. *Geophys. Mem. Dublin Inst. Adv. Stud.*, No. 2, Pt. 1.

— and MURPHY, T. 1952. Measurements of gravity in Ireland. Gravity survey of Ireland north of the line Sligo–Dundalk. *Geophys. Mem. Dublin Inst. Adv. Stud.*, No. 2, Pt. 4, pp. 1–36.

CONYBEARE, W. and BUCKLAND, W. 1816. Descriptive notes referring to the outline sections presented by a part of the coasts of Antrim and Derry. *Trans. Geol. Soc. London.* 1st Series, Vol. 3, pp. 196–216.

CRUICKSHANK, J. G. 1970. Soils and pedogenesis in the north of Ireland. P. 403 in *Irish Geographical Studies.* STEPHENS, N. and GLASSCOCK, R. W. (Editors). (Belfast: The Queen's University.)

CUMMINGS, G. L. and RICHARDS, J. R. 1975. Ore Lead Isotope Ratios in a continuously changing earth. *Earth Planet Sci. Let.*, Vol. 28, pp. 155–171.

DAVIS, J. W. 1890. Fossil fish-remains from Carboniferous shales at Cultra, County Down. Ireland. *Proc. Yorkshire Geol. Polytech Soc.*, Vol. 11, pp. 332–334.

DAY, J. B. W. 1970. Geology of the country around Bewcastle. *Mem. Geol. Surv. G.B.*

DE SITTER, L. 1956. *Structural Geology.* 552 pp. (London: McGraw Hill.)

DEWEY, J. F. 1961. A note concerning the age of the metamorphism of the Dalradian rocks of western Ireland. *Geol. Mag.*, Vol. 98, pp. 399–405.

— 1971. A model for the Lower Palaeozoic evolution of the southern margin of the early Caledonides of Scotland and Ireland. *Scott. J. Geol.*, Vol. 7, pp. 219–240.

DICKIE, G. 1859. On a deposit of Diatomaceae and Mollusca in the County of Antrim. *Q. J. Microgr. Sci.*, Vol. 7, pp. 9–11.

DIXON, F. E. 1949. Irish mean sea level. *Sci. Proc. R. Dublin Soc.*, Vol. 25, pp. 1–8.

DOYLE, J. B. 1853. Notes on the Salt Mine of Duncrue. *Trans. Geol. Soc. Dublin.*, Vol. 5, pp. 232–236 [fig. p. 231].

DUBOURDIEU, J. 1802. *Statistical survey of the County of Down.* Dublin.

DUNHAM, K. C. 1973. A recent deep borehole near Eyam, Derbyshire. *Nature, London*, Vol. 241, No. 108, pp. 84–85.

— and ROSE, W. C. G. 1949. Permo-Triassic geology of south Cumberland and Furness. *Proc. Geol. Assoc.*, Vol. 60, pp. 11–40 (also discussion by Stubblefield, C. J., pp. 39–40).

DU NOYER, G. V. 1868. On worked flint flakes from Carrickfergus and Larne. *Q. J. Geol. Soc. London*, Vol. 24, p. 495.

— 1869a. On the flint flakes of Antrim and Down. *J. R. Geol. Soc. Irel.*, Vol. 2, pp. 169–171.

— 1869b. On flint flakes from Carrickfergus and Larne. *Q. J. Geol. Soc. London*, Vol. 25, pp. 48–50.

DWERRYHOUSE, A. R. 1923. The glaciation of north-east Ireland. *Q. J. Geol. Soc. London.*, Vol. 79, pp. 352–422.

EMELEUS, C. H. and PRESTON, J. 1969. *Field excursion guide to the Tertiary volcanic rocks of Ireland.* 70 pp. (Belfast.)

EUNSON, H. J. 1884. The range of the Palaeozoic rocks beneath Northampton. *Q. J. Geol. Soc. London*, Vol. 40, pp. 482–496.

EYLES, V. A. 1952. The composition and origin of the Antrim laterites and bauxites. *Mem. Geol. Surv. North. Irel.*

FIRTH, W. A. and SWANSTON, W. 1887. Reference to the diatomaceous deposits at Lough Mourne and in the Mourne Mountains. *Proc. Belfast Nat. Field Club*, Series 2, Vol. 3, pp. 62–64.

FLETCHER, T. P. 1964. *In* The Centenary Year of the Belfast Naturalists' Field Club. *Ir. Nat. J.*, Vol. 14, pp. 221.

— 1967. *Correlation of the Cretaceous exposures of east Antrim.* Unpublished M.Sc. thesis, Queen's University, Belfast.

— 1978. Lithostratigraphy of the chalk (Ulster White Limestone Formation) in Northern Ireland. *Rep. Inst. Geol. Sci.*, No. 77/24.

— and WOOD, C. J. *in* WILSON, H. E. and MANNING, P. I. 1978. *q.v.*

FOLK, R. L. and PITTMAN, J. S. 1971. Length-slow chalcedony: a new testament for vanished evaporites. *J. Sediment. Petrol.*, Vol. 41, pp. 1045–1058.

FOWLER, A. 1944. A deep bore in the Cleveland Hills. *Geol. Mag.*, Vol. 81, pp. 193–206.

— 1955. The Permian of Grange, Co. Tyrone. *Bull. Geol. Surv. G.B.*, No. 8, pp. 38–53.

— 1959. The non-ferrous minerals of Northern Ireland. Pp. 27–34 in *The future of non-ferrous mining in Great Britain and Ireland.* (London: Institute of Mining and Metallurgy.)

— and ROBBIE, J. A. 1961. Geology of the country around Dungannon. *Mem. Geol. Surv. North. Irel.*

FRESHNEY, E. C. 1961. *The Cementstone Group of the West Midland Valley of Scotland.* Unpublished thesis, University of Glasgow.

GARWOOD, E. J. 1931. The Tuedian Beds of northern Cumberland and Roxburghshire east of the Liddel Water. *Q. J. Geol. Soc. London*, Vol. 87, pp. 97–159.

GEOLOGICAL SURVEY OF GREAT BRITAIN. 1895. *Annual Report of the Geological Survey and Museum of Practical Geology for year ending 31 Dec. 1894.* Appendix E, Part A, p. 288. (London: HMSO.)

— — — — 1959. *Summary of Progress of the Geological Survey of Great Britain and the Museum of Practical Geology for 1959.* (London: HMSO.)

GEOLOGICAL SURVEY OF NORTHERN IRELAND. 1967. Gravity anomaly map of Northern Ireland, 1:253,440 scale.

— — — — 1971. Magnetic anomaly map of Northern Ireland, 1:253,440 scale.

GEORGE, T. N. 1953. The Lower Carboniferous rocks of north-west Ireland. *Adv. Sci.*, pp. 65–73.

— 1958. Lower Carboniferous palaeogeography of the British Isles. *Proc. Yorkshire Geol. Soc.*, Vol. 31, pp. 227–318.

— 1960. The stratigraphical evolution of the Midland Valley. *Trans. Geol. Soc. Glasgow*, Vol. 24, pp. 32–107.

— 1967. Land form and structure in Northern Ireland. *Scot. J. Geol.*, Vol. 3, pp. 413–418.

— JOHNSON, G. A. L., MITCHELL, M., PRENTICE, J. E., RAMSBOTTOM, W. H. G., SEVASTOPULO, G. D. and WILSON, R. B. 1976. A correlation of Dinantian rocks in the British Isles. *Spec. Rep. Geol. Soc. London*, No. 7, 87 pp.

GOUGH, G. C. 1905–6. Foraminifera of the greensand of Whitehead. *Ir. Nat.*, Vol. 14, pp. 109.

GOVERNMENT OF NORTHERN IRELAND. 1925. Commission on the natural and industrial resources of Northern Ireland. *Report on the Mineral Resources of Northern Ireland.* 91 pp. (Belfast: HMSO.)

GRAINGER, J. 1859. On the shells found in the Post Tertiary deposits of Belfast. *Nat. Hist. Rev.*, Vol. 6, pp. 135–151.

GREEN, R. 1963. Lower Mississippian ostracods from the Banff Formation, Alberta. *Bull. Res. Counc. Alberta*, Vol. 11, pp. 1–237.

GRIFFITH, A. E. 1961. Note on some shelly fossils from the arenaceous greywackes of County Down. *Ir. Nat. J.*, Vol. 13, pp. 258–259.

— BAZLEY, R. A. B. and CAMERON, I. B. In prep. Geology of the country around Enniskillen. *Mem. Geol. Surv. North. Irel.*

GRIFFITH, R. 1837. Presidential Address. *J. Geol. Soc. Dublin*, Vol. 1, pp. 146–149.

— 1838. Second report of the commission appointed to consider and recommend a general system of railway for Ireland.

— 1843. On the lower portion of the Carboniferous Limestone Series of Ireland. *Rep. Brit. Assoc.*, pp. 45–46.

— 1861. Catalogue of the several localities in Ireland where mines of metalliferous indications have hitherto been discovered, arranged in counties, according to their respective post towns. *Dublin Q. J. Sci.*, Vol. 1, p. 244.

GUNN, P. J. 1973. Location of the proto-Atlantic suture in the British Isles. *Nature, London*, Vol. 242, pp. 111–112.

HANCOCK, J. M. 1961. The Cretaceous System in Northern Ireland. *Q. J. Geol. Soc. London*, Vol. 117, pp. 11–36.

— 1963. The hardness of the Irish Chalk. *Ir. Nat. J.*, Vol. 14, pp. 157–164.

HARDMAN, E. T. 1873. On analysis of White Chalk from the County of Tyrone, with a note on occurrence of zinc therein, and in the overlying basalt. *Geol. Mag.*, Vol. 10, pp. 434–438.

HARKER, A. 1904. The Tertiary igneous rocks of Skye. *Mem. Geol. Surv. Scot.*

HARRIS, 1774. The ancient and present state of the County of Down. 271 pp. (Dublin: Exshaw.)

HARTLEY, J. J. 1935. The underground water resources of Northern Ireland. *Inst. Civ. Eng. (N.I. Assoc.)*

— 1940. The sub-soils of Belfast and District. *Inst. Civ. Eng. (N.I. Assoc.)*

— 1943. Notes on the Lower Marls of the Lagan Valley. *Ir. Nat. J.*, Vol. 8, pp. 128–132.

— 1949. Further notes on the Permo-Triassic rocks of Northern Ireland. *Ir. Nat. J.*, Vol. 9, pp. 314–316.

— and HARPER, J. J. 1937. A recently discovered Ordovician inlier in County Down with a note on a new species of Pyritonema. *Ir. Nat. J.*, Vol. 6, pp. 253–255.

HAUGHTON, J. 1852. Account of the gangue of Conlig Lead Mine, County of Down. *J. Geol. Soc. Dublin*, Vol. 5, p. 203.

HENMI, C., KUSACHI, I., HENMI, K., SABINE, P. A. and YOUNG, B. R. 1973. A new mineral, bicchulite, the natural analogue of gehlenite hydrate, from Fuka, Okayama Prefecture, Japan, and Carneal, County Antrim, Northern Ireland. *Mineralog. J.* [Japan], Vol. 7, pp. 243–251.

HILL, A. R. and PRIOR, D. B. 1968. Directions of Ice Movement in north-east Ireland. *Proc. R. Ir. Acad.*, Vol. 66B, pp. 71–84.

HIRST, D. M. and DUNHAM, K. C. 1963. Chemistry and petrography of the Marl Slate of S. E. Durham, England. *Econ. Geol.*, Vol. 55, pp. 912–940.

HOME, H. 1912. Worked Flints obtained from 'the 25 foot Raised Beach' near Holywood, County Down. *Nature, London*, Vol. 90, p. 361.

HOSPERS, J. and CHARLESWORTH, H. A. K. 1954. The natural remanent of magnetisation of the Lower Basalts of Northern Ireland. *Monogr. Notes R. Astron. Soc. Geophys. Suppl.*, No. 7, pp. 32–43.

HUGHES, C. J. 1972. Spilites, keratophyres and the igneous spectrum. *Geol. Mag.*, Vol. 109, pp. 513–527.

HULL, E. 1873.   On the raised beach of the north-east of Ireland. *Rep. Brit. Assoc.* [for 1872], pp. 113–114.

— 1876.   The Country around Antrim, Larne and Carrickfergus. Explanatory Memoir to accompany Sheets 21, 28 and 29 of the Maps. *Mem. Geol. Surv. Irel.*

— WARREN, J. L. and LEONARD, W. B. 1871.   The country around Bangor, Newtownards, Comber and Saintfield in the County of Down. *Mem. Geol. Surv. Irel.*

HUME, W. F. 1897.   The Cretaceous strata of County Antrim. *Q. J. Geol. Soc. London*, Vol. 53, pp. 540–606.

HUTT, J. 1968.   A redescription of the Llandoverian monographed '*Graptolithus*' *tenuis* Portlock 1843. *Geol. Mag.*, Vol. 105, pp. 251–255.

INSTITUTE OF GEOLOGICAL SCIENCES. 1972.   Aeromagnetic map of Great Britain, Sheet 1, 1:625,000 scale.

— — — 1973.   *Annual Report for 1972.* (London: IGS.)

IVIMEY-COOK, H. C. 1975.   The stratigraphy of the Rhaetic and Lower Jurassic in East Antrim. *Bull. Geol. Surv. G.B.*, No. 50, pp. 51–70.

JESSEN, K. 1949.   Studies in late–Quaternary deposits and flora–history of Ireland. *Proc. R. Ir. Acad.*, Vol. 52B, pp. 85–290.

JONES, T. R. 1884.   Notes on the late Mr George Tate's specimens of Lower Carboniferous Entomostraca from Berwickshire and Northumberland. *Hist. Berwickshire Nat. Club*, Vol. 10, pp. 312–326.

— and KIRBY, J. W. 1885.   Notes on Palaeozoic bivalved Entomostraca. No. 19. On some Carboniferous species of the ostracodous genus *Kirkbya*. *Annu. Mag. Nat. Hist.*, (5), Vol. 15, pp. 174–190.

— — 1886a.   On Carboniferous Ostracoda from the Gayton Boring, Northamptonshire. *Geol. Mag.*, (3), Vol. 3(6), pp. 248–253.

— — 1886b.   Notes on the Palaeozoic bivalved Entomostraca. No. 22. On some undescribed species of British Carboniferous Ostracoda. *Annu. Mag. Nat. Hist.*, (5), Vol. 18, pp. 249–269.

— — 1886c.   On some fringed and other Ostracoda from the Carboniferous Series. *Geol. Mag.*, (3), Vol. 3, (10), pp. 433–439.

— — 1896.   On Carboniferous Ostracoda from Ireland. *Sci. Trans. R. Dublin Soc.*, (2), Vol. 6, pp. 173–200.

JOPE, E. M. 1957.   Carrickfergus Castle (official Illustrated Guide). (Belfast: HMSO.)

JUKES, J. B. 1868.   The Chalk of Antrim. *Geol. Mag.*, Vol. 5, pp. 345–347.

JUKES-BROWN, A. J. 1903.   The Cretaceous rocks of Britain. Vol. 2. *Mem. Geol. Surv. Eng. and Wales.*

KANE, R. 1845.   *The Industrial Resources of Ireland.* 2nd Edn. (Dublin).

KELLING, G. 1961.   The stratigraphy and structure of the Ordovician rocks of the Rhinns of Galloway. *Q. J. Geol. Soc. London*, Vol. 117, pp. 37–75.

KELLY, J. 1868.   On the geology of the County of Antrim with parts of the adjacent counties. *Proc. R. Ir. Acad.*, Vol. 10, pp. 235–321.

KINAHAN, G. H. 1885–89.   Economic geology of Ireland. *J. R. Geol. Soc. Irel.*, Vol. 8.

KING, W. 1852.   On the Permian fossils of Cultra. *Rep. Brit. Assoc.* [for 1852.]

— 1857.   On the occurrence of Magnesian Limestone at Tullyconnell near Ardtrea. *J. Geol. Soc. Dublin*, Vol. 7, pp. 67–81.

KNOWLES, W. J. 1880.   Flint implements from the raised beach at Larne and other parts of the north-east coast of Ireland. *J. R. Ir. Acad.*, Vol. 2, pp. 209–213.

— 1914.   The antiquity of man in Ireland, being an account of the older series of Irish flint implements. *R. Anthropol. Inst. J.*, Vol. 44, pp. 83–121.

LAMONT, A. 1946.   Red flints. *Ir. Nat. J.*, Vol. 8, pp. 398–399.

LAMPLUGH, G. W. and OTHERS. 1904.   The geology of the country around Belfast. *Mem. Geol. Surv. Irel.*

LAPWORTH, C. 1877.   On the graptolites of County Down. *Proc. Belfast Nat. Field Club*, App. IV (1876–77), pp. 125–147.

— 1878.   The Moffatt Series. *Q. J. Geol. Soc. London*, Vol. 34, pp. 240–346.

— 1880.   On new British graptolites. *Ann. Mag. Nat. Hist.*, Ser. 5, Vol. 5, pp. 149–177, pls. 4, 5.

LEAKE, B. E. 1963.   The location of the Southern Uplands Fault in Central Ireland. *Geol. Mag.*, Vol. 100, pp. 420–423.

LEEDER, M. R. 1973.   Lower Carboniferous serpulid patch reefs, bioherms and biostromes. *Nature, London*, Vol. 242, pp. 41–42.

LLOYD, H., SABINE, E. and ROSS, J. C. 1836.   Observations on the direction and intensity of the terrestrial magnetic force in Ireland. *Rep. Br. Assoc.* [for 1835], *5th Meeting (Dublin)*.

LOCKWOOD, F. W. 1884a.   On the recent examination of the Crannogs at Lough Mourne near Carrickfergus. *Proc. Belfast Nat. Field Club*, Series 2, Vol. II for 1882–83, pp. 170–174.

— 1884b.   The Crannogs of Lough Mourne [Abstract]. *J. R. Hist. Archeol. Assoc. of Irel.*, (4), Vol. 6, i (1883), pp. 177.

— 1884c.   Account of the Crannogs of Lough Mourne near Carrickfergus. *J. R. Hist. Archeol. Assoc. of Irel.*, (4), Vol. 6, i (1883), pp. 194–195.

MACDONALD, G. A. 1953.   Anhydrite-gypsum equilibrium relations. *Am. J. Sci.*, Vol. 251, pp. 884–898.

MACDONALD, R. and MACWILLIAMES, J. G. 1934.   A new Holocene deposit in County Down. *J. Conchol.*, Vol. 20, pp. 91–92.

— — 1938.   Raised beach section at Groomsport, County Down. *J. Conchol.*, Vol. 24, p. 103.

— — 1961.   A further raised beach section at Groomsport, County Down. *Ir. Nat. J.*, Vol. 13, p. 260.

MACADAM, J. 1850a.   Observations on the neighbourhood of Belfast with a description of cuttings on the Belfast and County Down railway. *J. Geol. Soc. Dublin*, Vol. 4, pp. 250–265.

— 1850b.   Supplementary observations on the neighbourhood of Belfast. *J. Geol. Soc. Dublin*, Vol. 4, pp. 265–268.

— 1852.   On the fossiliferous beds of the counties of Antrim and Down. *Rep. Br. Assoc.* [for 1852], p. 53.

MCALEESE, D. and MCCONAGHY, S. 1957–58.   Studies on the basaltic soils of Northern Ireland, *J. Soil Sci.*, Vol. 8, pp. 127–140, Vol. 9, pp. 66–88, 289–297.

MCLEAN, A. C. and QURESHI, I. R. 1966.   Regional gravity anomalies in the western Midland Valley of Scotland. *Trans. R. Soc. Edinburgh*, Vol. 66, pp. 267–283.

MCCONNELL, J. D. C. 1955.   The hydration of larnite (β-Ca$_2$ SiO$_4$) and bredigite (α-Ca$_2$ SiO$_4$) and its properties of the resulting gelatinous mineral plombierite. *Mineral. Mag.*, Vol. 30, pp. 672–680.

MCCRUM, E. J. 1909.   *A history of the antiquities of Carrickfergus.* (Belfast.)

MCGUGAN, A. 1957.   Upper Cretaceous foraminifera from Northern Ireland. *J. Palaeont.*, Vol. 31, pp. 329–348.

MCHENRY, A. and WATTS, W. W. 1898.   Guide to the collections of rocks and fossils belonging to the Geological Survey of Ireland.

McKerrow, W. S. 1959. The Southern Upland Fault in Ireland. *Geol. Mag.*, Vol. 96, pp. 347–352.

McMillan, N. F. 1938. A kitchen midden at Greenisland. *Ulster J. Archeol.*, (3), Vol. 1, pp. 16–19.

— 1947. The Estuarine Clay at Greenisland, Co. Antrim. *Ir. Nat. J.*, Vol. 9, pp. 16–19.

— 1949. Notes on Post-Glacial Clays in Anglesey. *Proc. Liverpool Geol. Soc.*, Vol. 20, pp. 106–110.

McQuiston, I. B. 1965. *Ostracods of Holocene deposits from the Belfast area, Northern Ireland.* MS Thesis, Urbana, Illinois.

Manning, P. I. 1960. New mineral records from County Down, Northern Ireland. *Ir. Nat. J.*, Vol. 13, p. 167.

— 1965. Visit to the Tennant Salt Mine, Eden, Carrickfergus. *Ir. Nat. J.*, Vol. 15, pp. 16–19.

— 1971. The development of water resources of Northern Ireland: progress towards integration. *Q. J. Eng. Geol.*, Vol. 4, pp. 335–353.

— Robbie, J. A. and Wilson, H. E. 1970. Geology of Belfast and the Lagan Valley. *Mem. Geol. Surv. North. Irel.*

— and Wilson, H. E. 1975. The stratigraphy of the Larne Borehole, County Antrim. *Bull. Geol. Surv. G.B.*, No. 50, pp. 1–50.

M'Skimin, S. 1811. *History of Carrickfergus.* (Belfast.)

Millar, J. 1780–1783. Journal of the progress of the mine works of the Bangor and Newtown Company. Public Records Office, Northern Ireland, D 654/P/1.

Miscampbell, A. 1894. The salt industry of Carrickfergus. *Trans. Fed. Inst. Miner. Eng.*, Vol. 7, pp. 546–552.

Mitchell, G. F. 1951. Studies in Irish Quaternary deposits. *Proc. R. Ir. Acad.*, Vol. 53B, pp. 111–206.

Mitchell, G. H. and Mykura, W. 1962. The geology of the neighbourhood of Edinburgh. *Mem. Geol. Surv. G.B.*

Mitchell, W. I. (in press). The geology of the country around Armagh. *Mem. Geol. Surv. North. Irel.*

Moorbath, S. 1962. Lead isotope abundance determination studies on mineral occurrences in the British Isles and their geological significance. *Philos. Trans. R. Soc. London*, Series A, Vol. 254, pp. 295–360.

Moore, E. 1861. On the zones of the Lower Lias and the *Avicula contorta* Zone. *Q. J. Geol. Soc. London*, Vol. 17, pp. 483–516.

Moore, R. C. 1961. Treatise on Invertebrate Palaeontology, Part Q *Anthrapoda 3, Crustacea, Ostracoda. Geol. Soc. Am. and Univ. Kansas Press.*

Morris, P. 1972. The magnetisation of a baked limestone contact in the County Mayo, Ireland. *Geophys. J. R. Astron. Soc.*, Vol. 29, pp. 487–492.

Movius, H. L. 1953a. Graphic representation of post-glacial changes of level in north-east Ireland. *Am. J. Sci.*, Vol. 251, pp. 697–740.

— 1953b. Curran Point, Larne, Co. Antrim. The type site of the Irish Mesolithic. *Proc. R. Ir. Acad.*, Vol. 56B, pp. 1–195.

Muir, I. D. and Tilley, C. E. 1961. Mugearites and their places in alkali igneous rock series. *J. Geol.*, Vol. 69, pp. 186–203.

Murchison, R. I. 1854. *Siluria. The history of the oldest known rocks containing organic remains with a brief sketch of the distribution of gold over the earth.* (London: Murray.)

Murphy, T. 1952. Measurements of gravity in Ireland. Gravity survey of central Ireland. *Geophys. Mem. Dublin Inst. Adv. Stud.*, No. 2, Pt. 3.

— 1953. The magnetic survey of Ireland for the Epoch 1950.5. *Geophys. Mem. Dublin Inst. Adv. Stud.*, No. 4.

— 1955. A vertical force magnetic survey of the counties Roscommon, Longford, Westmeath and Meath. *Geophys. Bull. Dublin Inst. Adv. Stud.*, No. 11.

Ogniben, L. 1957. Petrografia della Serie Solifera Siciliana e considerazioni geologiche relative. *Mem. Carta Geol. Ital.*, Vol. 33, 268 pp.

Oliver, G. J. H. 1978. Prehnite-pumpellyite facies metamorphism in County Cavan, Ireland. *Nature, London*, Vol. 274, pp. 242–243.

Orme, A. R. 1966. Quaternary changes in sea-level in Ireland. *Trans. Inst. Br. Geogr.*, Vol. 39, pp. 127–140.

Patterson, E. M. 1941. Caledonian and Tertiary dykes in County Down. *Ir. Nat. J.*, Vol. 7, pp. 333–335.

— 1950. Evidence of fissure eruption in the Tertiary Lava Plateau of north-east Ireland. *Geol. Mag.*, Vol. 87, pp. 45–52.

— 1951a. An occurrence of quartz-trachyte among the Tertiary basalt lavas of north-east Ireland. *Proc. R. Ir. Acad.*, Vol. 53B, pp. 265–287.

— 1951b. Notes on three Tertiary dolerites from Counties Down and Antrim. *Ir. Nat. J.*, Vol. 10, pp. 178–181.

— 1952. A petrochemical study of the Tertiary lavas of north-east Ireland. *Geochim. Geophys. Acta.*, Vol. 2, pp. 283–299.

— and Swaine, D. J. 1957. The Tertiary dolerite plugs of north-east Ireland. A survey of their geology and geochemistry. *Trans. R. Soc. Edinburgh*, Vol. 63, pp. 317–331.

Patterson, L. I. 1892. Newly discovered site for the working of flint in the County of Down. *J. R. Soc. Antiq. Irel.*, Vol. 22, pp. 154–155.

Pattison, J. 1970. A review of the marine fossils from the Upper Permian rocks of Northern Ireland and north-west England. *Bull. Geol. Surv. G.B.*, No. 32, pp. 123–165.

Peach, B. N. 1901. On a remarkable volcanic vent of Tertiary age in the Island of Arran, enclosing Mesozoic fossiliferous rocks. *Q. J. Geol. Soc., London*, Vol. 57, pp. 126–143.

Peake, N. B. and Hancock, J. M. 1961. The Upper Cretaceous of Norfolk. *Trans. Norfolk Norwich Nat. Soc.*, Vol. 19, pp. 293–339.

Pollock, J. and Wilson, H. E. 1961. A new fossiliferous locality in County Down. *Ir. Nat. J.*, Vol. 13, pp. 244–248.

Portlock, J. E. 1843. *Report on the geology of the County of Londonderry and parts of Tyrone and Fermanagh.* 784 pp. (Dublin: HMSO.)

Powell, D. W. 1970. Magnetised rocks within the Lewisian of western Scotland and under the Southern Uplands. *Scott. J. Geol.*, Vol. 6, pp. 353–369.

Praeger, R. L. 1890. Report of a committee of investigation on the gravels and associate beds of the Curran, at Larne, County Antrim. *Rep. Proc. Belfast Nat. Field Club*, Series 2, Vol. III (1889–1890), pp. 198–209.

— 1892. Report on the estuarine clays of the north-east of Ireland. *Proc. R. Ir. Acad.*, (3), Vol. 2, pp. 212–289.

— 1897. Report upon the raised beaches of the north-east of Ireland, with special reference to their fauna. *Proc. R. Ir. Acad.*, (3), Vol. 4, pp. 30–54.

— 1902. Summary of the Post-Glacial sequence in the Belfast district. *Br. Assoc. Adv. Sci. Rep.*, pp. 611–612.

— and Coffey, G. 1904. The Antrim Raised Beach: A contribution to the Neolithic history of the north of Ireland. *Proc. R. Ir. Acad.*, Vol. 25, pp. 143–200.

Preston, J. 1971. The Maddygalla Dyke, Rathlin Island. *Ir. Nat. J.*, Vol. 17, pp. 88–92.

Prior, D. B. 1966. Late-Glacial and post-Glacial shorelines in north-east Antrim. *Ir. Geogr*, Vol. 5, pp. 173–185.

Ramsbottom, W. H. C. 1977. Major cycles of transgression and regression (mesotherms) in the Namurian. *Proc. Yorkshire Geol. Soc.*, Vol. 41, pp. 261–291.

REED, F. R. C. 1952. Revision of certain Ordovician fossils from County Down. *Proc. R. Ir. Acad.*, Vol. 55B, pp. 29–136.

REID, R. E. H. 1958. Remarks on the Upper Cretaceous Hexactinellida of County Antrim. *Ir. Nat. J.*, Vol. 12, pp. 236–243, 261–269.

— 1962. The Cretaceous succession in the area between Red Bay and Garron Point, Co. Antrim. *Ir. Nat. J.*, Vol. 14, pp. 73–77.

— 1963. New records of *Gonioteuthis* in Ireland. *Ir. Nat. J.*, Vol. 14, p. 98.

— 1964. The Lower (pre-*Belemnitella mucronata*) White Limestone of the east and north-east of Co. Antrim. *Ir. Nat. J.*, Vol. 14, pp. 262–297, 296–303.

— 1971. The Cretaceous rocks of north-eastern Ireland. *Ir. Nat. J.*, Vol. 17, pp. 105–129.

REMANE, A. 1934. Die brackwasserfauna (mit besonderer Berücksichtigung der Ostsee). *Verh. dt. zool. Gea.*, Supple., Vol. 7, pp. 34–74.

REYNOLDS, D. L. 1928. The petrography of the Triassic sandstone of north-east Ireland. *Geol. Mag.*, Vol. 65, pp. 448–473.

— 1931. The dykes of the Ards Peninsula. *Geol. Mag.*, Vol. 68, pp. 97–111, 145–165.

RICHARDSON, W. 1803. Accounts of whynne dykes in the neighbourhood of the Giant's Causeway, Ballycastle, and Belfast. *Trans. R. Ir. Acad.*, Vol. 9, pp. 21–43.

RICKARDS, R. B. and RUSHTON, A. W. A. 1968. The thecal form of some slender Llandovery *Monograptus*. *Geol. Mag.*, Vol. 105, pp. 264–274.

RIGBY, J. 1905. Outburst from Duncrue Old Salt Mine after being tapped for brine. *Trans. Manchester Geol. Min. Soc.*, Vol. 28, pp. 565–570.

RÜCKER, A. W. and THORPE, T. E. 1891. A magnetic survey of the British Isles for the Epoch January 1, 1886. *Philos. Trans. R. Soc.*, Series A, Vol. 181, pp. 53–328.

— — 1896. A magnetic survey of the British Isles for the Epoch January 1, 1891. *Philos. Trans. R. Soc.*, Series A, Vol. 188, pp. 1–661.

SABINE, E. 1870. Contribution to terrestrial magnetism No. XII. The magnetic survey of the British Isles reduced to Epoch 1842–5. *Philos. Trans. R. Soc.*, Series A, Vol. 160, pp. 265–275.

SABINE, P. A. 1968. Ferrian chlorospinel from Carneal, Co. Antrim. With chemical analysis by G. A. Sergeant, and X-ray data by B. R. Young. *Mineral. Mag.*, Vol. 36, pp. 948–954.

— 1975. Refringence of iron-rich wollastonite. *Bull. Geol. Surv. G.B.*, No. 52, pp. 65–67.

— STYLES, M. T. and YOUNG, B. R. 1982. Gehlenite, an exomorphic mineral from Carneal, Co. Antrim, Northern Ireland. *Rep. Inst. Geol. Sci.*, No. 82/1, pp. 61–63.

— and YOUNG, B. R. 1975. Metamorphic processes at high temperature and low pressure: the petrogenesis of the metasomatised and assimilated rocks of Carneal, Co. Antrim. *Philos. Trans. R. Soc.*, Series A, Vol. 280, pp. 225–269.

SANDERS, I. S. and MORRIS, J. H. 1978. Evidence for Caledonian subduction from greywacke detritus in the Longford–Down inlier. *J. Earth Sci. R. Dublin Soc.*, Vol. 1, pp. 53–62.

SCHARFF, R. F. 1903. Animal remains from the Gobbins Caves, Co. Antrim. *Ir. Nat.*, Vol. 12, pp. 55–56.

SCHNELLMANN, G. A. 1964–65. Recent developments in the search for minerals in the United Kingdom. *Report of Joint Meeting with the Institution of Mining Engineers. Trans. Inst. Min. Metal.*, Vol. 74, Pt. 2.

SEYMOUR, H. J. 1899. In *Summary of Progress of the Geological Survey of Great Britain and the Museum of Practical Geology for 1898*. Pp. 180–181.

SHARPE, E. N. 1970. An occurrence of Pillow Lavas in the Ordovician of County Down. *Ir. Nat. J.*, Vol. 16, pp. 299–301.

SHERIDAN, D. J. R. 1972. Upper Old Red Sandstone and Lower Carboniferous of the Slieve Beagh Syncline and its setting in the northwest Carboniferous basin, Ireland. *Spec. Pap. Geol. Surv. Irel.*, No. 2.

— HUBBARD, W. F. and OLDROYD, R. W. 1967. A note on Tournaisian strata in Northern Ireland. *Sci. Proc. R. Dublin Soc.*, Vol. 3A, pp. 33–37.

SHERLOCK, R. L. 1924. British Salt Deposits. *Rep. Br. Assoc.* [for 1923], pp. 442–443. (See also The Quarry, Vol. 29, pp. 38–41, 171.)

— 1926. A correlation of the British Permo-Triassic rocks, Pt. 1, North England, Scotland and Ireland. *Proc. Geol. Assoc.*, Vol. 37, pp. 1–72.

— 1928. A correlation of the British Permo-Triassic rocks, Pt. 2. England south of the Pennines and Wales. *Proc. Geol. Assoc.*, Vol. 39, pp. 49–95.

SIMPSON, L. 1963. The stratigraphy and tectonics of the Manx Slate Series, Isle of Man. *Q. J. Geol. Soc. London*, Vol. 119, pp. 368–396.

SINGH, G. and SMITH, A. G. 1966. The Post-Glacial marine transgression in Northern Ireland—conclusions from estuarine and raised beach deposits: a contrast. *Palaeobotanist*, Vol. 15, pp. 230–234.

SMITH, D. B. 1970. The palaeogeography of the British Zechstein. Pp. 20–23 in *Third Symposium on Salt*, Vol. 1, DELWIG, L. F. and RAU, J. L. (Editors). (Cleveland: Northern Ohio Geological Society.)

— BRUNSTROM, R. G. W., MANNING, P. I., SIMPSON, S. and SHOTTON, F. W. 1974. A correlation of Permian rocks in the British Isles. *J. Geol. Soc. London*, Vol. 130, pp. 1–45.

SMITH, D. I. 1961. Patterns of minor faults in the south central Highlands of Scotland. *Bull. Geol. Surv. G.B.*, No. 17, pp. 145–152.

SMITH, G. F. H., ASHCROFT, F. N. and PRIOR, G. T. 1916. Chabazite and associated minerals from County Antrim. *Mineral. Mag.*, Vol. 17, pp. 274–304.

SMITH, J. D. D. 1957. Graptolites with associated sedimentary grooving. *Geol. Mag.*, Vol. 94, pp. 425–428.

SMITH, W. 1850. On deposits of diatomaceous earth found on the shores of Lough Mourne, Co. Antrim, with a record of species living in the lake. *Ann. Mag. Nat. Hist.* (N.S.) V.

STACEY, J. S. and KRAMERS, J. D. 1975. Approximation of terrestial lead isotope evolution by a two stage model. *Earth Planet Sci. Let.*, Vol. 26, pp. 207–221.

STAPLES, J. H. 1869. The flaked, chipped and worked flints to be found in the gravel in the neighbourhood of Holywood, County Down. *6th Annu. Rep. Proc. Belfast Nat. Field Club*, p. 42.

STEPHENS, N. 1957. Some observations on the "Interglacial" Platform and the Early Post-Glacial Raised Beach on the East Coast of Ireland. *Proc. R. Ir. Acad.*, Vol. 58B, pp. 129–149.

— 1958. The Evolution of the coast-line of north-east Ireland. *Adv. Sci.*, Vol. 56, pp. 389–391.

— 1963. Late-glacial sea-levels in north-east Ireland. *Geography*, Vol. 4, pp. 345–359.

— 1965. Geomorphological Map of north-east Ireland. *The Geology of the Irish Sea Area. Second Colloquium.* (Dublin: University College.)

— and COLLINS, A. 1960. The Quaternary Deposits at Ringneill Quay and Ardmillan, Co. Down. *Proc. R. Ir. Acad.*, Vol. 61C, pp. 41–77.

— and SYNGE, F. M. 1966. Pleistocene shorelines. Pp. 1–51 in *Geomorphological essays*. DRURY, G. H. (Editor). (London: Heinemann.)

STEVENS, R. E. 1944. Composition of some chromites of the Western Hemisphere. *Am. Mineral.*, Vol. 29, pp. 1–34.

STEWART, D. 1800. The Report of Donald Stewart, Itinerant Mineralogist to the Dublin Society. *Trans. Dublin Soc.*, Vol. 1.

STEWART, S. A. 1871a. A list of fossils of the Estuarine Clays of the Counties of Down and Antrim. *Proc. Belfast Nat. Field Club*, App. II, No. 2, pp. 27–42.

— 1871b. The latest fluctuations of the sea-level on our own coasts. *8th Annu. Rep. Proc. Belfast Nat. Field Club*, pp. 55–57.

— 1871c. A list of the fossils of the Estuarine clays of Co. Down and Co. Antrim. *8th Annu. Rep. Belfast Nat. Field Club* (for 1870–71) Appendix pp. 27–40.

— 1879–80. Mollusca of the Boulder Clay in north-east Ireland. *Proc. Belfast Nat. Field Club*, pp. 165–176.

STIRRUP, M. 1877. The Raised Beaches of County Antrim, their molluscan fauna and flint implements. *Proc. Lit. Philos. Soc. Manchester*, Vol. 16.

STRACHAN, J. 1907–8. The origin and formation of zeolites in basalt. *Proc. Belfast Nat. Field Club*, Vol. 2 (6), p. 92.

SWANSTON, W. 1877. On the Silurian Rocks of Co. Down, Pt. 1, Correlation. *Proc. Belfast Nat. Field Club*, App. IV (1876–7), pp. 107–123.

— 1886. Report of the committee appointed to investigate the Larne gravels, and to determine the position in them of the flint, flakes and cores for which they are noted. *Annu. Rep. Proc. Belfast Nat. Field Club*, Series 2, Vol. 2, pp. 519–530.

SYNGE, F. M. and STEPHENS, N. 1960. The Quaternary Period in Ireland—An Assessment. *Ir. Geogr.*, Vol. 4, pp. 121–130.

— — 1966. Late and Post-Glacial shorelines and ice limits in Argyll and North-East Ulster. *Trans. Inst. Br. Geogr.*, Vol. 39, pp. 101–145.

TATE, R. 1864. On the Liassic strata of the neighbourhood of Belfast. *Q. J. Geol. Soc. London*, Vol. 20, pp. 103–114.

— 1865. On the correlation of the Cretaceous formations of the north-east of Ireland. *Q. J. Geol. Soc. London*, Vol. 21, pp. 15–44.

— 1867. On the Lower Lias of the north-east of Ireland. *Q. J. Geol. Soc. London*, Vol. 23, pp. 297–305.

— 1870. Note on the Middle Lias in the north-east of Ireland. *Q. J. Geol. Soc. London*, Vol. 26, pp. 324–325.

TAYLOR, B. J., PRICE, R. H. and TROTTER, F. M. 1963. Geology of the country around Stockport and Knutsford. *Mem. Geol. Surv. G.B.*

TILLEY, C. E. 1952. Some trends of basaltic magma in limestone syntexis. *Am. J. Sci.*, Bowen Vol., pp. 529–545.

— and HARWOOD, H. F. 1931. The dolerite-chalk contact of Scawt Hill, Co. Antrim. The production of basic alkali-rocks by the assimilation of limestone by basaltic magma. *Mineral. Mag.*, Vol. 22, pp. 439–468.

THIRLAWAY, H. I. S. 1951. Measurements of gravity in Ireland. Gravity observations between Dublin, Sligo, Galway and Cork. *Geophys. Mem. Dublin Inst. Adv. Stud.*, No. 2, Pt. 2.

THOMPSON, S. M. 1894. Section at Greenisland Station in Report of the Geological Committee for 1893–1894. *Proc. Belfast Nat. Field Club.*, Vol. 2, (4), pp. 104–127.

TOGHILL, P. 1968a. A new Lower Llandovery graptolite from Coalpit Bay, Co. Down. *Geol. Mag.*, Vol. 105, pp. 384–386.

— 1968b. The graptolite assemblages and zones of the Birkhill Shales (Lower Silurian) at Dobb's Linn. *Palaeontology*, Vol. 11, pp. 654–668.

TOMKEIEFF, S. I. 1934. Differentiation in basalt lava, Islandmagee, Co. Antrim. *Geol. Mag.*, Vol. 71, pp. 501–512.

TURNER, F. J. 1936. Metamorphism of the Te Anau Series in the region north-west of Lake Wakatipu. *Trans. R. Soc. N.Z.*, Vol. 65, pp. 329–349.

TURNER, J. S. 1950. Notes on the Carboniferous Limestone of Ravenstonedale, Westmorland. *Trans. Leeds Geol. Assoc.*, Vol. 6, (3), pp. 124–134.

— 1952. The Lower Carboniferous Rocks of Ireland. *Liverpool Manchester Geol. J.*, Vol. 1, pp. 113–147.

— and MARSHALL, C. E. 1940. The Killough–Ardglass dyke swarm. *Q. J. Geol. Soc. London*, Vol. 96, pp. 321–338.

VACQUIER, V., STEENLAND, N. C., HENDERSON, R. G. and ZIETZ, I. 1951. Interpretation of Aeromagnetic Maps. *Mem. Geol. Soc. Am.*, Vol. 47.

VERNON, P. 1966. Drumlins and Pleistocene ice-flow over the Ards Peninsula/Strangford Lough area, County Down, Ireland. *J. Glaciol.*, Vol. 5, pp. 401–409.

WALKER, G. P. L. 1959a. Some observations on the Antrim Basalts and associated dolerite intrusions. *Proc. Geol. Assoc.*, Vol. 70, pp. 179–205.

— 1959b. The amygdale minerals in the Tertiary lavas of Ireland. II. The distribution of gmelinite. *Mineral. Mag.*, Vol. 32, pp. 202–217.

— 1960a. The amygdale minerals in the Tertiary lavas of Ireland. III. Regional distribution. *Mineral. Mag.*, Vol. 32, pp. 503–527.

— 1960b. An occurrence of Mugearite in Antrim. *Geol. Mag.*, Vol. 97, pp. 62–64.

— 1962a. A note on occurrences of tree remains within the Antrim Basalts. *Proc. Geol. Assoc.*, Vol. 73, pp. 1–7.

— 1962b. Low-potash gismondine from Ireland and Iceland. *Mineral. Mag.*, Vol. 33, pp. 187–201.

— 1962c. Garronite, a new zeolite from Ireland and Iceland. *Mineral. Mag.*, Vol. 33, pp. 173–186.

WALKER, G. W. 1919. The magnetic re-survey of the British Isles for the Epoch January 1, 1915. *Philos. Trans. R. Soc.*, Series A, Vol. 219, pp. 1–72.

WARRINGTON, G. and AUDLEY-CHARLES, M. G., ELLIOTT, R. E., EVANS, W. B., IVIMEY-COOKE, H. C., KENT, P. E., ROBINSON, PAMELA, L., SKOTTON, F. W. and TAYLOR, F. M. 1980. A correlation of the Triassic Rocks in the British Isles. *Spec. Rep. Geol. Soc. London*, No. 13, 78 pp.

WATTS, W. A. 1962. Early Tertiary pollen deposits in Ireland. *Nature, London*, Vol. 193, pp. 600–601.

— 1963. Fossil Seeds from the Lough Neagh Clays. *Ir. Nat. J.*, Vol. 14, pp. 117–118.

WEAVER, T. 1825. Report to the Hibernian Mining Company. National Library, Dublin MS 658.

WELCH, R. J. 1902. The Gobbins Cliffs and Caves, Co. Antrim. *Ir. Nat.*, Vol. 11, pp. 214–216.

— 1904. Greensand section at Whitehead. *Ir. Nat.*, Vol. 13, p. 49.

— 1914. Estuarine Clay section at Holywood. *Ir. Nat.*, Vol. 23, pp. 239–240.

WELLS, A. K. 1936. [Report of the Committee on] Petrographic Nomenclature. *Geol. Mag.*, Vol. 73, pp. 319–325.

WELCH, R. J. 1902. The Gobbins cliffs and caves. *Ir. Nat.*, Vol. 11, pp. 214–216.

— 1904. Greensand section at Whitehead. *Ir. Nat.*, Vol. 13, p. 49.

WEST, I. M. 1964. Evaporite diagenesis in the Lower Purbeck Beds of Dorset. *Proc. Yorkshire, Geol. Soc.*, Vol. 34, pp. 315–330.

— 1965. Macrocell structure and enterolithic veins in British Purbeck gypsum and anhydrite. *Proc. Yorkshire Geol. Soc.*, Vol. 35, pp. 47–58.

— 1973. Vanished evaporites—significance of strontium minerals. *J. Sediment. Petrol.*, Vol. 43, pp. 278–279.

— BRANDON, A. and SMITH, M. 1968. A tidal flat evaporite facies in the Visean of Ireland. *J. Sediment. Petrol.*, Vol. 38, pp. 1079–1093.

WHELAN, C. B. 1928. The implementiferous raised beach gravels of south-east Antrim. *Man*, Vol. 28, pp. 186–189.

WILLCOX, N. R. 1955. Some faecal pellets from Cretaceous strata in Co. Antrim. *Ir. Nat. J.*, Vol. 11, pp. 265–271.

WILLIAMS, A. 1959. A structural history of the Girvan district, S. W. Ayrshire. *Trans. R. Soc. Edinburgh*, Vol. 63, pp. 629–667.

WILSON, H. E. 1972. *The Regional Geology of Northern Ireland.* 113 pp. (Belfast: HMSO.)

— and ROBBIE, J. A. 1966. Geology of the country around Ballycastle. *Mem. Geol. Surv. North. Irel.*

— and MANNING, P. I. 1978. Geology of the Causeway Coast. *Mem. Geol. Surv. North. Irel.*

WILSON, R. L. 1959. Remanent magnetism of late secondary and early Tertiary British rocks. *Philos. Mag.*, Vol. 4, pp. 750–755.

— 1961. Palaeomagnetism in Northern Ireland. *Geophys. J. R. Astron. Soc.*, Vol. 5, pp. 45–69.

— 1970. Palaeomagnetic stratigraphy of Tertiary lavas from Northern Ireland. *Geophys. J. R. Astron. Soc.*, Vol. 20, pp. 1–9.

— and SMITH, P. J. 1968. The nature of secondary natural magnetizations in some igneous baked rocks. *J. Geomagn. Geoelectr.*, Vol. 20, pp. 367–380.

WOLFE, M. J. 1968. Lithification of a carbonate mud: Senonian Chalk in Northern Ireland. *Sediment. Geol.*, Vol. 2 (4), pp. 263–290.

— 1968. An electron-microscope study of the surface texture of sand grains from a basal conglomerate. *Sedimentology*, Vol. 8, pp. 2139–2147.

WOODROW, A. 1978. A history of the Conlig and Whitespots Lead Mines. British Mining No. 7, *Mon. North. Mine Res. Soc.*

WRIGHT, J. 1872. Geology of Cultra, Co. Down. *Annu. Rep. Belfast Nat. Field Club (1871–72)*, pp. 33–37.

— 1880. Post-tertiary foraminifera of north-east Ireland. *Annu. Rep. Proc. Belfast Nat. Field Club*, (2), Vol. 1, pp. 428–429 and A list of the Post-tertiary foraminifera of the north-east of Ireland. *ibid* Appendix 5, pp. 149–164.

— 1880. Recent foraminifera of Down and Antrim. *Annu. Rep. Proc. Belfast Nat. Field Club*, Systematic lists illustrative of the Flora, Fauna, Palaeontology and Archaeology of the North of Ireland. Appendix 4, pp. 101–106.

— 1911. Foraminifera from the Estuarine Clays of Magheramorne, Co. Antrim, and Limavady, Co. Londonderry. *Proc. Belfast Nat. Field Club* for 1910–11, Appendix (No. II of Vol. III).

— 1902. The Marine Fauna of the Boulder Clay. *Rep. Brit. Assoc.*

WRIGHT, J. E., HULL, J. H., McQUILLIN, R. and ARNOLD, SUSAN E. 1971. Irish Sea investigations 1969–70. *Rep. Inst. Geol. Sci.*, No. 71/19, 55 pp.

WRIGHT, W. B. 1919. An analysis of the Palaeozoic floor of north-east Ireland, with predictions as to concealed coalfields. *Sci. Proc. R. Dublin Soc.*, Vol. 15, (NS), pp. 629–650.

— in COMMISSION ON THE NATURAL AND INDUSTRIAL RESOURCES OF NORTHERN IRELAND. 1925. *Report on the mineral resources of Northern Ireland.* (Belfast: Government of Northern Ireland.)

YOUNG, R. 1873. Some remarks on recent changes of coast level at Ballyholme Bay, County Down. *Proc. Belfast Nat. Hist. Philos. Soc.* for 1871–2, pp. 39–41.

LIST OF GEOLOGICAL SURVEY PHOTOGRAPHS

Copies of these photographs are deposited for public reference in the Library of the Institute of Geological Sciences, South Kensington, London SW7 2DE and in the office of the Geological Survey of Northern Ireland, 20 College Gardens, Belfast BT9 6BS. Prints and lantern slides are supplied at a fixed tariff on application to the Director. These photographs, which belong to Series NI, were taken mainly by J. M. Pulsford with a few by H. E. Wilson.

346 Flow-banded trachyte, 300 m ENE of Carneal Bridge.
347 General view of the Carneal dolerite plug. Alluvial flat in foreground.
348 Reaction rims round flint nodules embedded in metamorphosed Chalk, Carneal plug.
349 Concentric reaction zone around a flint nodule embedded in metamorphosed Chalk, Carneal plug.
350 Dry gorge in Cretaceous White Limestone, Altfraechan Glen, Redhall, Ballycarry.
351 General view of Larne Lough and Island Magee. Oldmill Bay.
352 General view of Larne Lough and Island Magee from Ballycarry Station.
353 Coastal landslip topography, Hillhead, Island Magee.
354 Coastal view, Cloghfin, Island Magee.
355 Quarry in White Limestone, Ballycarry Limeworks.
356 General view of south-end of Larne Lough, north-west of Muldersleigh Hill.
357 Upper Cretaceous (Hibernian Greensand) outcrop, Cloghfin, Island Magee.
358 Upper Cretaceous (Hibernian Greensand), Cloghfin, Island Magee.
359 Post-Senonian sandstone in solution channels in White Limestone, south of Cloghfin, Island Magee.
360 Clay-with-flints in solution hollow in Chalk, south of Cloghfin, Island Magee.
361 'The Lord,' Sea-stack of massive Lower Basalt, Island Magee.
362 Dyke intruded into Lower Basalt, Boating Hole, Island Magee.
363 Volcanic agglomerate of Lower Basalt age, north of Black Head, Island Magee.
364 Curved pipe amygdales in a lava flow, north of Black Head, Island Magee.
365 Amygdale pipes in a lava flow, north of Black Head, Island Magee.
366 Lower Basalt lava flows, Black Head, Island Magee.
367 Lower Basalt lava flows, White Head, Island Magee.
368 Columnar structure in the Lower Basalts; railway cutting south of White Head Quarry.
369 General view of southern Island Magee and Whitehead.
370 Lower Basalt escarpment, Knockagh, Greenisland.
371 Overflow channel in Lower Basalt escarpment, Knockagh, Greenisland.
372 Penarth Group shales, Woodburn Glen North, Carrickfergus.
373 Upper Cretaceous Hibernian Greensand, Woodburn Glen North, Carrickfergus.
374 Upper Cretaceous Hibernian Greensand detail, Woodburn Glen North, Carrickfergus.
375 Disused brickpit in Keuper Marl, Carrickfergus.

376–8 Carrickfergus Castle.
379 Thermal alteration of Keuper Marl by Tertiary dolerite dyke, Carrickfergus.
380 Greenisland Sill, intruded into Keuper Marl, Greenisland.
381–5 Salt beds at the Tennant Mine, Eden, Carrickfergus, Co. Antrim.
386 Faulted junction between Carboniferous and Permian strata, Cultra.
387 Unconformity of Trias (Bunter Marl) on Permian, Cultra.
388 Carboniferous strata of the Ballycultra Formation, Cultra.
389–90 Algal limestones Ballycultra Formation, Cultra.
391 Shales faulted against grits (Ordovician) Cultra.
392 Cleavage in grit, Craigavad.
393 Sedimentary structures on sole of bedding plane (Ordovician) Craigavad.
394 Worm casts on bedding plane (Ordovician) Craigavad.
395 Asymmetrical anticline in Ordovician rocks, shore 400 m NE of Craigavad.
396 Erratic boulders, shore 480 m NE of Craigavad.
397 Nodule in shale, shore section 400 m NE of Craigdarragh.
398 Lithology of Ordovician rocks, shore section 400 m NE of Craigdarragh.
399 Drumlin in Upper Boulder 300 m WSW of Craigdarragh.
400 Boulder clay, south bank of Railway cutting at Glencraig.
401 Ordovician lithology, Holywood.
402 Bedding and cleavage, shore section 200 m NE of Carnalea Station.
403 Strain slip cleavage and parallel quartz veins, on shore 400 m NE of Carnalea Burn, Co. Down.
404 Raised beach, west side of Smelt Mill Bay.
405 Glossy shales, 150 m N of Pickie Pool.
406 Graded bedding 300 m SW of Luke's Point.
407 Typical lithology of Ordovician rocks, Luke's Point.
408 Twisted regional cleavage, south of Orlock Tunnel.
409 Post-cleavage microfolds 300 m N of Orlock Bridge.
410 Inverted current bedding 150 m ESE of Orlock Bridge.
411 Concretionary nodules 150 m ESE of Orlock Bridge.
412 Portavo Group—argillaceous lithology. Cow and Calf rocks.
413 Raised beach escarpment on Copeland Island seen from Cow and Calf rocks.
414 Morainic debris, Ballast Pit north of disused railway line, Hogstown.
415 Donaghadee Sandstone Group, Robbys Point.
416 Crumpled Ordovician shales, Coalpit Bay, Donaghadee.
417 Kinnegar Grits, headland south of Galloways Bridge, Donaghadee.
418 Groove-casts on sole of massive greywacke grit, 150 m N of Galloways Bridge.
419 Groove-casts on sole of massive greywacke grit, 150 m N of Galloways Bridge.
420 Post-cleavage dome, due east of Galloways Bridge.

421–2 Sole-markings on greywacke grits, 120 m S of Galloways Bridge.

423 Feather joints, 125 m S of Galloways Bridge.

424 Inclusion in greywacke sandstone, 125 m S of Galloways Bridge.

425 Glacial sands and gravel, Knocknagoney Stream.

426 Asymmetrical anticline in Silurian beds, Ballykeel Quarry.

427 Feature along Conlig lead lode, South Engine Shaft, Conlig Lead Mines.

428 Breccia from lead lode. Tip heaps at North Engine Shaft, Conlig Lead Mines.

429 Moraine, Killaghy 1 km SW of Ballyvester School.

430 Morainic sand and gravel. Eagle Hill, Craigboy.

431 Platform of peat on modern beach, Templepatrick, Millisle.

432 Sinistral wrench fault, Kinnegar Rocks.

433 Texture of Kinnegar Grits, Kinnegar Rocks.

633 Claypit in slipped Triassic mudstones, Knockagh.